GLOBAL CLIMATE CHANGE

Linking Energy, Environment, Economy, and Equity

ENVIRONMENTAL SCIENCE RESEARCH

Series Editor:

Herbert S. Rosenkranz
Department of Environmental and Occupational Health
Graduate School of Public Health
University of Pittsburgh
130 DeSoto Street
Pittsburgh, Pennsylvania

Founding Editor:
Alexander Hollaender

Recent Volumes in this Series

Volume 36 – SHORT-TERM BIOASSAYS IN THE ANALYSIS OF COMPLEX ENVIRONMENTAL MIXTURES V
Edited by Shahbeg S. Sandhu, David M. DeMarini, Marc J. Mass, Martha M. Moore, and Judy L. Mumford

Volume 37 – HAZARDS, DECONTAMINATION, AND REPLACEMENT OF PCB: A Comprehensive Guide
Edited by Jean-Pierre Crine

Volume 38 – *IN SITU* EVALUATION OF BIOLOGICAL HAZARDS OF ENVIRONMENTAL POLLUTANTS
Edited by Shahbeg S. Sandhu, William R. Lower, Frederick J. de Serres, William A. Suk, and Raymond R. Tice

Volume 39 – GENETIC TOXICOLOGY OF COMPLEX MIXTURES
Edited by Michael D. Waters, F. Bernard Daniel, Joellen Lewtas, Martha M. Moore, and Stephen Nesnow

Volume 40 – NITROARENES: Occurrence, Metabolism, and Biological Impact
Edited by Paul C. Howard, Stephen S. Hecht, and Frederick A. Beland

Volume 41 – ENVIRONMENTAL BIOTECHNOLOGY FOR WASTE TREATMENT
Edited by Gary S. Sayler, Robert Fox, and James W. Blackburn

Volume 42 – CHEMISTRY FOR THE PROTECTION OF THE ENVIRONMENT
Edited by L. Pawlowski, W. J. Lacy, and J. J. Dlugosz

Volume 43 – PRIMARY PRODUCTIVITY AND BIOGEOCHEMICAL CYCLES IN THE SEA
Edited by Paul G. Falkowski and Avril D. Woodhead

Volume 44 – SECONDARY-METABOLITE BIOSYNTHESIS AND METABOLISM
Edited by Richard J. Petroski and Susan P. McCormick

Volume 45 – GLOBAL CLIMATE CHANGE: Linking Energy, Environment, Economy, and Equity
Edited by James C. White

GLOBAL CLIMATE CHANGE
Linking Energy, Environment, Economy, and Equity

Edited by

James C. White

Professor Emeritus
Cornell University
Ithaca, New York

Associate editors

William R. Wagner and Carole N. Beal

Center for Environmental Information
Rochester, New York

PLENUM PRESS • NEW YORK AND LONDON

Library of Congress Cataloging in Publication Data

Global climate change: linking energy, environment, economy, and equity / edited by James C. White; associate editors, William R. Wagner and Carole N. Beal.
p. cm.—(Environmental science research; v. 45)
"Proceedings of the eighth annual conference of the Air Resources Information Clearinghouse on Global Climate Change: Linking Energy, Environment, Economy, and Equity, held December 5-6, 1991, in Washington, D.C."—T.p. verso.
Includes bibliographical references and index.

1. Energy development—Environmental aspects—Congresses. 2. Climatic changes—Congresses. 3. Air quality management—Congresses. 4. Energy policy—Congresses. I. White, James C. (James Carrick), 1916- II. Air Resources Information Clearinghouse (U.S.) III. Series.
TD195.E49G57 1992 92-26925
363.7—dc20 CIP

Proceedings of the Eighth Annual Conference of the Air Resources Information Clearinghouse on Global Climate Change: Linking Energy, Environment, Economy, and Equity, held December 5–6, 1991, in Washington, D.C.

ISBN 978-1-4419-3229-7

A Division of Plenum Publishing Corporation
233 Spring Street, New York, N.Y. 10013

Printed in the United States of America

PREFACE

The nations of the world are not facing up to global environmental needs. We are failing to recognize the potential severity of approaching problems and are not investing enough in prevention and amelioration to stem the deterioration of our environment.

We face burgeoning population growth and the attendant further increase in energy demand. With our current technology, nearly all that energy will come from fossil fuels and further strain our systems.

Environmental damage is a cost attached to fossil fuel energy production and is largely an unpaid expense, but we must make a start at paying those costs if we are to leave future generations an environmentally sound world.

The threat of climate change has brought nations together in an unprecedented way to negotiate a world-wide agreement, to stabilize and reduce greenhouse gases. However not all nations are ready to commit to specific timetables and targets. There are many skeptics of present policies, and little real faith that we will be able to meet and hold the goals we propose today.

If an agreement is to be realized more than scientific data must be considered. Energy needs must be linked with environmental damage, with economic development and with the rights of less developed countries to approach equality in energy use.

In an effort to spell out these linkages, The Center for Environmental Information (CEI) organized a meeting in December, 1991, titled "Global Climate Change: Linking Energy, Environment, Economy and Ethics." The program was organized by a distinguished steering committee representing many disciplines and constituencies, and was sponsored by more than thirty organizations.

This book is a compilation of the papers from that meeting and would not have been possible without the cooperative efforts of the staff of CEI. We appreciate the work of associate editors William Wagner and Carole Beal, in helping to bring this volume to print, and Debra Segura is to be thanked for her capable work in preparing the manuscript.

The papers in this volume bring together the supply and the demand sides of energy issues and gives approaches to integrating energy, economic and social concerns with the environment. They present strategies for decision-making on local, national and international levels, and contain two innovative articles on the ethics of living in a global greenhouse. The presentations attempt to mold the thoughts from diverse fields concerned with climate change, and a special value lies in the discussions of what policies might work in facing up to a changing world.

This cross-disciplinary approach should make the book useful to scholars in many areas concerned with global change.

James C. White
Cornell University, 1992

ABOUT THE CENTER FOR ENVIRONMENTAL INFORMATION

The Center for Environmental Information (CEI) was established in Rochester, New York, in 1974 as an answer to the growing dilemma of where to find timely, accurate and comprehensive information on environmental issues. To meet this need for current and comprehensive information, CEI has developed a multi-faceted program of publications, educational programs and information services. It is a private, nonprofit organization funded by membership dues, fees, contracts, grants and contributions.

The Center remains today a Rochester-based organization, but its services now reach far beyond the local community, reflecting the increasing number, scope and complexity of problems affecting the environment.

CEI acts as a catalyst to advance the public agenda toward soundly conceived environmental policies. CEI's communication network provides a link among the scientific community, educators, decision makers and the public, so that informed action follows the free interchange of information and ideas.

CONFERENCE STEERING COMMITTEE

Priscilla Auchincloss, Congressional Fellow, U.S. House of Representatives
Jan Beyea, Senior Staff Scientist, National Audubon Society
Alex Cristofaro, Director, Air & Energy Policy Division, U.S. Environmental Protection Agency
Roger C. Dower, Director, Climate, Energy and Pollution Program, World Resources Institute
Michael P. Farrell, Section Head, Earth Systems, Oak Ridge National Laboratory
T.J. Glauthier, Director, Energy and Climate Change, World Wildlife Fund and The Conservation Foundation
Marcel Halberstadt, Director, Environmental Department, Motor Vehicle Manufacturers Association of the United States, Inc.
Nelson E. Hay, Chief Economist and Director, Policy Analysis, American Gas Association
Richard Richels, Senior Program Manager, Environmental Risk Analysis, Electric Power Research Institute
James C. White, Center for the Environment, Cornell University
Richard Wilson, Director, Northeastern Regional Center for Global Environmental Change, Lyman Laboratory of Physics, Harvard University

CONFERENCE COSPONSORS

Funding Cosponsors:

American Gas Association
Electric Power Research Institute
National Oceanic and Atmospheric Administration
Oak Ridge National Laboratory, Center for Global Environmental Studies
U.S. Environmental Protection Agency
World Resources Institute

Contributing Cosponsors:

American Petroleum Institute
E.I. DuPont de Nemours & Co., Inc.
Edison Electric Institute
Motor Vehicle Manufacturers Association of the United States, Inc.
Ontario Ministry of the Environment
World Wildlife Fund and The Conservation Foundation

Cooperating Cosponsors:

Alliance to Save Energy
American Council for an Energy Efficient Economy
Center for Global Change, University of Maryland
Center for the Environment, Cornell University
Climate Institute
Environment Canada
Environmental and Energy Study Institute
Friends of the Earth, Environmental Policy Institute
Izaac Walton League of America
National Audubon Society
National Institute for Emerging Technology
National Science Foundation
National Wildlife Federation
Natural Resources Defense Council
Quebec Ministere de l'Environnement
Union of Concerned Scientists
United Nations Environment Programme
U.S. Council for Energy Awareness
U.S. Department of Agriculture, Economic Research Service
World Meteorological Organization

CONTENTS

Opening Remarks 1
Elizabeth Thorndike

Challenges in the Face of Uncertainty 3
Ray T. Oglesby

Discussion 9

A SUSTAINABLE ENERGY FUTURE - TECHNOLOGICAL OPTIONS FOR CHANGE

Steps to Reduce U.S. Carbon Dioxide Emissions: Technical Options and the Policies to Implement Them 13
Robert M. Friedman and Rosina M. Bierbaum

Energy Efficiency, Human Activities and Climate Change 37
Lee Schipper

The Costs of Reducing U.S. CO_2 Emissions: Further Sensitivity Analyses 45
Alan S. Manne and Richard G. Richels

Discussion 53

INTEGRATING ENERGY, ECONOMY, AND ENVIRONMENTAL CONCERNS

"No Regrets" Tax Reform 61
Alex Cristofaro

Multiple Benefit Environmental Policies 65
Daniel Lashof

Energy and Environmental Policy: The Role of Markets 73
Daniel J. Dudek

Local/Regional Policies - Acting Globally by Thinking Locally 81
Edward L. Hillsman

Discussion 91

IMPLICATIONS FOR DECISION MAKING

Sectoral Perspectives

Government Perspectives 103
Howard Gruenspecht

The Electric Utilities' Perspective 107
William L. Fang

Auto Industry Perspectives 111
Samuel A. Leonard

Industrial Sector Perspectives 115
Frederick E. Mangelsdorf

Discussion 121

Societal Considerations

The Importance of Population Growth in Future Commercial Energy Consumption 127
Gretchen Kolsrud and Barbara Boyle Torrey

The Public As Decision Maker 143
Rob Coppock

Discussion 147

STRATEGIES FOR INTERNATIONAL NEGOTIATIONS

Status of Framework Convention Talks 153
Daniel A. Reifsnyder

Equity in International Agreements 157
Richard Stewart

Discussion 161

THE ETHICS OF LIVING IN A GLOBAL GREENHOUSE

Corporate and Personal Responsibility 169
Dale Jamieson

Intergenerational Equity 177
Douglas MacLean

Discussion 183

EXAMINING THE LINKAGES - WHICH POLICIES WILL WORK?

Energy Policy: Insurance Against the Unknown 189
David Bodansky

Designing a Carbon Tax Strategy 195
Roger Dower

"Virtual Climate" and Climate Change Assessment: Paving the Way for Workable Climate Change Policies 203
Charles N. Herrick

Examining the Linkages - Which Policies Will Work? 207
Toufiq Ali

A Congressional View of Workable Linkages 211
John D. Dingell

Discussion 215

Thinking About Climate Change 223
Milton Russell

Conference Participants 229

Conference Program 233

Index 237

OPENING REMARKS

Elizabeth Thorndike

President
Center for Environmental Information, Inc.
46 Prince Street
Rochester, NY 14607

I am Elizabeth Thorndike, president of the Center for Environmental Information, and I want to welcome you to this eighth annual conference of the Center.

Important to me, and I think probably important to many of you here, is the fact that the conference steering committee worked very hard to include a large number of women on this agenda. You don't see them. I think we have only two women on the program. It was not for lack of trying, but I think it is an indication that when the original choices of women are not available, there aren't other sources of options--an interesting message for those of you concerned about these things. However, there are a number of very outstanding women involved with the global climate change issue.

This climate change conference is the first, to our knowledge, which has chosen to make connections between the natural sciences, the social sciences and underlying ethical considerations. Societal realities such as population growth, the increasingly powerful role of the public as decision maker, questions of equity among nations and between generations, must consciously become a part of the policy maker's deliberations. These considerations are of equal importance to those regarding scientific findings and economic impacts of specific mitigation or adaptation strategies.

Global climate change needs to be seen for what it is – the ultimate global risky business. As the burgeoning field of risk communication research tells us, qualitative as well as quantitative aspects of the issue need to be communicated, debated and evaluated if we are ever to arrive at a global consensus on how best to address the problem. This conference is intended as a step in that direction.

Global Climate Change, Edited by J.C. White
Plenum Press, New York, 1992

CHALLENGES IN THE FACE OF UNCERTAINTY

Ray T. Oglesby

Interim Director
Global Environment Program
Cornell University
Ithaca, New York 14850

This conference is focused on action, on ways to adapt to or control large-scale climate change. Before becoming too involved with such specific issues, however, it might be wise to pause and take an overview of the context within which we will be working. What are the assumptions underlying the need for action? What are the impediments to acceptance of these needs by policymakers? I shall begin with a consideration of the second question.

Uncertainty in all long-range climate predictions is frequently cited as a reason for doing little or nothing in the way of adaptation or control. Enough is known about the climate-affecting potential of unpredictable events, such as massive volcanic eruptions, to insure that some uncertainty will always be present. Add to this the sheer complexity of climate prediction mechanisms and the many determinants of climate still poorly understood and a prima facie case for inaction does not seem unreasonable. However, it would be highly imprudent, if not irresponsible, to accept this argument. On the contrary, there is an urgent need for rapid, comprehensive action to meet the many challenges posed by climate predictions—even in their current state of imperfection.

Why is such an activist approach justified? First, there are trends towards major changes in future climate which can be predicted with reasonable certainty. These trends are unidirectional (unless some major changes are made in the way humans carry out their functions on this planet), rapid and of such a nature as to pose serious threats to humankind. Second, planning and management horizons for much of our physical and economic infrastructure and for the management of natural resources such as forests and water stretch over timeframes of many decades. If climate change is not taken into account when planning and setting policies for these, the inevitable stresses produced by climate change will be greatly magnified.

What then are the aspects of future climate about which we can be "reasonably certain"? To answer this question we must begin with a consideration of the so-called

Global Climate Change, Edited by J.C. White
Plenum Press, New York, 1992

greenhouse gases. It has been established that these are accumulating with unprecedented rapidity in Earth's atmosphere. Because of their heat absorptive and storage properties, warming of the lower atmosphere is assured. As the world's population continues to grow and as people's aspirations continue to be driven by the ideals of western technology, accelerated inputs of greenhouse gases, with concomitant effects on planetary temperature, can be expected to continue for at least the next several decades. All of the above suppositions are strongly supported by the international scientific community.

There are three other virtual certainties that are important but which have received insufficient attention. One is that Earth's warming will be greater at higher than at lower latitudes. Very roughly, a mean global surface increase in temperature of 2.5 degrees Celsius translates into a 1-degree increase in equatorial temperature and as much as 10-12 degrees at the polar circles. The resultant global temperature pattern is one of much greater spatial uniformity. This has profound implications for Earth's major wind and ocean current systems and for the location, extent and productivity of biotic communities.

A second highly likely property of climate change is that warming will occur to a greater extent in winter than in summer with a resulting decrease in temperature difference between seasons. At first glance this "tropicalization" of temperate and high latitude climates may seen like a good thing, but it poses some very real problems. For example, municipal water supplies in those regions that depend upon mountain snowpacks for their water storage function may have a difficult time finding appropriate and affordable substitutes.

A third virtual certainty of the climate change scenario concerns linkage of all other aspects of climate to temperature. The nature and implication of some linkages is readily apparent (an example will be given later on); others are not yet in clear focus.

To return to the question of uncertainties, what aspects of future climate pose particular problems regarding their predictability? These fall into three categories—those related to temporal and spatial scales, those concerning aspects of climate other than atmospheric temperature and those dependent upon accurate forecasts of greenhouse gas concentrations in the atmosphere. The first relates to the way in which climate-predicting models are constructed and, in particular, limitations imposed on them by computing capability. This is a complex problem to understand and will not be addressed here. Suffice it to say that current scales are much coarser than those needed to view climate change in the contexts within which most planning activities and policy decisions are typically made.

Important aspects of climate other than temperature about which we are eager to have a predictive capability include patterns of storminess, rates of evaporation relative to precipitation and wind regimes. Current prediction of these climate properties contain many uncertainties. Nevertheless, in the case of some of these climate properties, educated guesses can be made about the nature of the changes, if not about their precise scalar values. For example, latitudinal temperature difference provides the force creating Earth's major wind regimes. With a predicted lessening of latitudinal temperature gradient, average velocity of the wind in prevailing systems will decrease. The implication for developing wind as a substitute for fossil fuel in the generation of electrical power is obvious.

The third major problem area relating to global climate change predictability concerns uncertainties in predicting greenhouse gas concentrations. Such estimates are made by solving mass balance equations where concentration change over a time interval equals what is there initially plus input over the time interval minus removal.

The latter depends upon chemical reactions in the atmosphere, outgassing to the stratosphere and then to space and deposition back onto Earth's surface. These processes are fairly well understood. Inputs depend upon human activities that can be summarized by the equation "input equals the size of Earth's population times the amount of technology used per person times some factor that expresses emission of radiatively active gases associated with each type of technology." Prediction of any one element of this equation is obviously fraught with uncertainty. About the best that can be done is to assume that present trends of population growth and technological use will continue on their present track for at least a few more decades. Climate modelers acknowledge and deal with uncertainties in greenhouse gas prediction by solving their models for low (radical shifts in technology used to decrease emissions), medium (business as usual) and high (accelerated emissions due to increased technology use by the Third World) scenarios of greenhouse gas inputs to the atmosphere.

To round out our overview of certainty and uncertainty in viewing climate change, it is appropriate that we briefly scan the mechanisms that provide us with estimates of future global climates. The principal tools used for such prediction are global climate models (GCMs). These consist of some twenty to thirty simultaneous differential equations operated with data input from both ground stations and earth-orbiting satellites processed by state-of-the-art supercomputers. Despite their apparent sophistication, GCMs are still able to only roughly simulate Earth's climate, partially due to technological and data limitations and in part to incomplete understanding of the climate system. At least six GCMs have been developed by independently operating groups of researchers, four of which are located in the United States. So long as support for GCM development continues, steady improvements, but not dramatic breakthroughs, can be anticipated. Remember, however, these tools, impressive as they are, will never give a perfect view of the future; uncertainties and downright surprises will always be there to challenge the researcher, planner and policymaker.

In addition to the atmospheric physicist's mathematical models, new insights into Earth's climate and its causes have come in recent years from studies of the past. For example, data from cores of the polar icecaps have demonstrated a close correlation between concentrations of the greenhouse gases carbon dioxide and methane in fossil air and atmospheric temperature at the time when these gases were trapped in ice. Other studies have shed light on the relation of ocean level to global temperature.

When GCM output and historical data analysis are considered together, a powerful case for Earth being in the early stages of a period characterized by rapid increases in global temperature accompanied by a suite of related changes in other climate parameters emerges. International and national review panels composed of highly qualified scientists who are not directly affiliated with the climate predicting efforts have concurred in this view.

To sit back and review the material so far presented leaves one with feelings akin to those produced by viewing a half-baked cake. The ingredients are known, the processes driving cake making are recognized and broadly understood, but one is not sure how palatable the final product may be nor when the cake will be ready for consumption. Nevertheless, it is certain the baking process will continue (barring some catastrophe like failure of power to the oven) and that an eating of the cake will take place in the near future. Uncertainty exists but is considered unimportant because knowledge and experience indicate a likely outcome of this exercise in cake baking.

Now I shall turn to another aspect of the global climate change scenario we should consider before immersing ourselves in the details of how to react. This is the question of perspective.

As humans we have taken great trouble to isolate ourselves from the unpleasant and dangerous aspects of climate; this is especially true for those of us who live in the technologically advanced societies. But this has been accomplished in a situation where climate has remained relatively constant. To consider what effects a unidirectional, rapid and universal trend of change might induce we must first face up to the problem of perspective. The majority of us, including those specifically charged with making plans and policies, operate on local and regional scales of space and years on a scale of time. Unfortunately, these are scales which our climate predicting tools are ill-equipped to handle. Within this small-scale framework we are left, as a result, with sets of directional arrows showing the probable nature of changes but only crude, uncertain estimates of their magnitude and timing.

Even these rough estimates can, in some cases, deliver unmistakable messages to the planner and policymaker. To take an example from an area with which I have some familiarity, water resources management, consider the problem of supplying water to the nine million plus inhabitants of New York City and the four counties to its north and west. Under normal operating conditions 90% of the 1.6 billion gallons per day supply is drawn from an intricate, beautifully engineered web of impoundments and aqueducts harvesting water collected as runoff from the Catskill Mountains. What will happen to this water supply if climate changes as currently predicted for the next half century? Summers will presumably be longer, hotter and drier, and winters shorter, warmer and wetter. Winter precipitation will run off the land more rapidly and earlier in the spring since more of it will fall as rain rather than as snow and what does arrive in the latter form will melt sooner. Additional impoundment capacity will be needed if this water is to be captured for later use by consumers. During the summer, demand will increase at the same time as replenishment decreases. The system already operates at near (some would maintain over) its capacity of supply. The ability of supply to meet demand becomes progressively more problematical with each year our hypothesized climate trend continues.

Municipal water resource planning typically takes place on a time scale of about a half century. Given projected climate changes over the next fifty years, with all their uncertainties, and given the present status of system capacity relative to demand, any prudent resident of the New York metropolitan area would expect public servants charged with providing them with water to plan for the contingency of climate change. I think one can safely assume that if the public was fully aware of the situation described above it would demand action, in the form of planning at a minimum, from its water authority.

The example of climate change and New York's water supply has taken us from cause (changing climate) to effect. It is the effects, not climate change per se, that spur us to seek means for adaptation, mitigation and, in the long run, control. In coming to grips with the problem of defining and evaluating effects there are two fundamental problems we should be aware of—establishing a sound approach and dealing with the multidisciplinary nature of almost all posited effects.

Two approaches are possible to gauging potential effects of climate change. These are construction of a deterministic model or extrapolation of historical analogues. The former is possible only when cause-and-effect relationships can be identified and defined. Some elements of a climate response scenario, e.g. purely physical phenomena such as evaporation rate, may yield to the first of these approaches. However, since virtually any response element in some way influences and is influenced by other elements, the model quickly becomes complex. Model robustness and accuracy is undermined as less well understood properties are included. The second approach, extrapolation of historical analogues, will likely be based on only a few examples and these will not exactly mimic the situation we wish to portray. Historical analogues probably offer a preferred way to estimate social responses to climate change and economic costs. Whichever approach is employed, we need to be fully aware of its potential, its limitations and the resources that will be needed for its implementation.

From the time global climate change scenarios are established, through estimation of their effects and, finally, to formulation of adaptation and control strategies the problem is a multidisciplinary one. The term "multidisciplinary" has taken on the status of a "buzzword" in recent years. Seldom is it defined in terms of specifics, and few who use it seem conscious of the difficulties in putting "multidisciplinary" into practice. But it must be done. Boundaries representing academic disciplines and agency responsibilities and authority, political and economic barriers, an almost pathological adherence by the research establishment to short-term (typically one to three years), narrowly defined topics—all must give way in the promotion of a multidisciplinary approach. Encouragingly, a few science administrators and a few politicians have recognized the need for multidisciplinary research into the global climate change and its effects scenario. Their belief has found its way into the structure of national and international research planning. Implementation is another thing. Actual disbursement of research funds is still largely according to established agency priorities with little attention to synthesis and minimal interchange between agencies. New perspectives, enlightened leadership and a solid funding base are needed if the many disciplines capable of contributing to such research are to be brought together in a meaningful context. This pervasive problem of conceptualizing and carrying out truly multidisciplinary research on national and international bases should receive very high priority from the research community.

Let us now consider some of the specific challenges facing those of us working together at this conference. The first relates to the problem of scale. I have already noted that, by virtue of training, intellectual and emotional inclination and professional responsibilities, each of us focuses on climate change and its effects at different points on the temporal and spatial scales. Acknowledgment that we are dealing with a global scenario unfolding over a period of decades is often overshadowed by a much narrower, highly focused viewpoint. Herein lies one of the principal challenges for us. We must constantly flex our minds to move between various scales of space and time.

Another challenge I lay before you is to be constantly aware of how disciplines, agencies and the private sector interact. Where are the roadblocks and the clear intersections between them? What techniques seem to work best in bridging gaps and exploring disagreements? Make a conscious effort to think and talk about such questions and you have played a positive role in promoting multidisciplinary research.

I shall conclude with a final thought. Those of us in this room and those who choose to read the conference proceedings represent an elite by virtue of education, resources at our command, and in the opportunities and abilities we have to influence government, academe and the private sector at all levels of organization and authority. This eliteness, taken together with the knowledge we share concerning Earth's future climate and its potential consequences for humankind and the natural world, imposes an awesome responsibility on us. If we and a small number of others who share these characteristics with us do not succeed in arousing widespread awareness and in developing responsible policies to deal with climate change and its effects, no one will, at least not until it is too late.

DISCUSSION

CHALLENGES IN THE FACE OF UNCERTAINTY

QUESTION: You depict warming as leading to water shortfalls for New York City. Why not also mention and consider that warming yields more evaporation yields; there will be more clouds and, therefore, more rainfall?

OGLESBY: Certainly many climate scientists recognize these possibilities. I was taking a scenario not as a forecast—that is certainly beyond my capabilities. My answer derives from two versions of global climate models.

Also, I borrowed from an EPA-financed study of the Feather River Watershed in California where the precipitation on an annual basis actually increased, but the water supply for the San Francisco Bay Area decreased because much more of the precipitation in the Sierras occurred in the winter and it came down as rainfall rather than snowfall. The system does not have the storage capacity to harvest that water and store it. It resulted in a wetter, warmer winter and a hotter, drier summer, and the entire system of storage and transportation of water was not predicated on that kind of scenario. The result was estimated to be a major shortfall in water supply for the Bay Area.

I maintain that there is enough validity in such scenarios that they should be played out routinely by water scientists, foresters, and many other people.

QUESTION: How does one explain the historical temperature record, particularly that in the United States? It does not correlate well with increases in greenhouse gas concentration.

OGLESBY: The world temperature pattern has been reviewed by people at the British Meteorological Office, Global Climate Study Center in East Anglia, and by Jim Hanson and his group at the Goddard Institute. They are the only two reviews of global temperatures that I have seen that seem to have been very carefully done. They show an increase in temperature of about a half degree Celsius since the industrial revolution went into full-scale operation about a hundred years ago. The models would predict somewhat more of an increase in temperature than that, but it is certainly within reasonable error estimates, insofar as the models are concerned.

The fact that the United States temperature may or may not have increased is a very difficult question to answer. I have been looking at some long-term phenological data sets for the northeastern United States, and they certainly suggest that, over the last hundred years, things like ice break-up on lakes had been occurring earlier, suggesting an onset of warmer weather earlier in the year.

There are all kinds of conflicting pieces of information out there. Part of the problem is that nobody collected information with the expectation that we were going to reconstruct a national or global climate later on, and any of you who are scientists know that, if you try to use data which was collected for one purpose for another, it is never entirely satisfactory. A second problem is how these data are interpreted, how much they are corrected for heat-island effects from urban centers and all of that sort of thing.

I don't know really what the United States record shows. I know it is a bone of contention among scientists, so I assume that the evidence is not conclusive one way or the other. That doesn't surprise me. The temperature record is very much biased in terms of geographical distribution of temperature collection stations, and we have a very short history of reliable temperature data in this country.

A few years back, W.H. Schlesinger and others did a study where they tried to get long-term historical record sets that were all comparable for their goals, and they wound up with only 54 points where they could say that temperature had been reliably defined for, I think, over 100 years.

We don't have that much information. I would maintain we really don't know what is happening. I think we will get at this by historical studies and other means and get a clearer picture over the next decade. But it wouldn't surprise me if glaciers were expanding in some parts of the world. This is because global average can translate into some pretty funny waves when it starts affecting ocean currents, for example.

QUESTION: Please repeat your viewpoint on temperature change in lower versus higher parts of the atmosphere and over what parts of the world.

OGLESBY: Let me see if I can explain this. The atmosphere around the earth consists of a very thin envelope of what we call air, and it is thicker at the equator than it is at the poles. It is cooler at the poles and cooler air is denser and tends to be compressed. This is a pretty simplistic explanation but I hope it will do. That means that, if we put a given quantity of heat into the atmosphere at the equator, it is put into a certain mass. If we put that same amount of heat into the atmosphere at higher latitudes, the mass is decreased because the layer is thinner and that means that proportionally it will warm more at higher latitudes than at lower latitudes. The same explanation holds for the winter/summer change.

I have seen statements which I have avoided because I think that they may be appropriate but they are almost too sensational to swallow.

For example, if the earth's climate responds as predicted over the next half century, at the end of that period the earth will be warmer than it has been in the last 125,000 years. In fact, the last time it was that warm, there were crocodiles at the Arctic Circle. That is a perhaps a rather overdramatic way of putting it, but it does indicate the order of magnitude that we are talking about.

A Sustainable Energy Future—Technological Options for Change

STEPS TO REDUCE U.S. CARBON DIOXIDE EMISSIONS:

TECHNICAL OPTIONS AND THE POLICIES TO IMPLEMENT THEM

Robert M. Friedman
Senior Associate

Rosina M. Bierbaum
Senior Associate and Project Director

Oceans and Environment Program
Office of Technology Assessment
U.S. Congress
Washington, DC 20510

INTRODUCTION

Among individual countries, the United States is the leading contributor of greenhouse gases. With 5 percent of the world's population, the United States accounts for about 20 percent of the world's warming commitment. Carbon dioxide (CO_2) is the most important greenhouse gas; U.S. carbon dioxide emissions (also about 20 percent of the global total) originate almost exclusively from fossil fuel combustion.

Many industrialized countries, principally in Europe, have called for either a freeze or a 20-percent reduction in CO_2 emissions from the developed world by 2000 or shortly thereafter; several have pledged to freeze or reduce emissions whether or not the rest of the world participates. Six committees of the Congress requested that the Office of Technology Assessment evaluate the potential for reductions in CO_2 emissions in the United States. This paper summarizes the results of that study, *Changing by Degrees: Steps to Reduce Greenhouse Gas Emissions.*[1]

Our study attempted two goals: (1) a detailed technical analysis estimating CO_2 emissions reduction potential in the U.S.; and (2) a detailed menu of policy options to control CO_2. There is no single right way to pursue emissions reductions. We offer some guidance, based on historical information and the kinds of energy users that would need to be influenced in each sector, and provide Congress with a variety of options.

Global Climate Change, Edited by J.C. White
Plenum Press, New York, 1992

In our study, we identified two packages of energy conservation, energy supply, and forest management practices to decrease CO_2 emissions over 25 years. We examined technical and policy measures that could be adopted in four sectors of the U.S. economy—buildings (commercial and residential), manufacturing, transportation, and energy supply. We modeled about 100 discrete, energy-related technical options, including relatively quick retrofits to existing equipment and, in the long term, introduction of more efficient technologies to replace worn out equipment and satisfy a growing economy's needs. The OTA model is based on a larger system of energy and economic models used by the Gas Research Institute to forecast energy use through 2010. OTA also identified steps to increase carbon uptake by U.S. forests and thereby offset CO_2 emissions, as well as methods to control emissions of other greenhouse gases, primarily from the food sector.

The most important conclusions of the OTA study include the following:

— U.S. CO_2 emissions will continue to grow for the foreseeable future in the absence of any new laws or other constructive efforts to reduce them. We estimate that CO_2 emissions will likely rise another 50 percent over the next 25 years (Figure 1).

— By adopting a package of low cost measures crossing all sectors (e.g., OTA's moderate scenario), the United States can significantly slow the growth of CO_2 emissions over the next 25 years. No single option nor focus on a single sector of the economy is sufficient.

— With more aggressive initiatives (e.g., OTA's tough scenario), it is conceivable that the United States could decrease its CO_2 emissions to 35 percent below today's levels by 2015. The reductions achievable within each sector of the economy for both OTA's moderate and tough scenarios are shown in Figure 1.

— A variety of federal policy initiatives will be required to achieve large reductions in U.S. CO_2 emissions. Such policy actions will have to include both regulatory "push" and market "pull" mechanisms, including performance standards, tax incentive programs, carbon-emission or energy taxes, labeling and efficiency ratings, and research, development, and demonstration activities. To achieve lasting reductions, government signals (e.g., pricing and regulatory policies) need to be consistent and reinforcing.

— CO_2 emissions are projected to rise again after 2015, even under OTA's tough scenario, unless vigorous and sustained research, development and demonstration (RD&D) programs for nonfossil energy supply alternatives (such as solar and nuclear power) are begun today—programs that can lead to commercialization of one or more of these alternatives by that time.

Overall, OTA believes that the moderate case is zero or very low-cost. Adding up the upper- and lower-cost bounds of the options we modeled suggests that our tough scenario could either save the United States a few tens of billions of dollars per year, cost up to $150 billion per year, or wind up somewhere in between. The range of uncertainty is so large because these are net compliance costs—the difference between two even larger numbers: capital and operating costs minus fuel savings. We estimate that annualized capital and operating costs will fall in the range of about $350 to $520 billion per year. However, fuel savings will be about $370 billion per year, assuming our projected 2015 fuel prices.

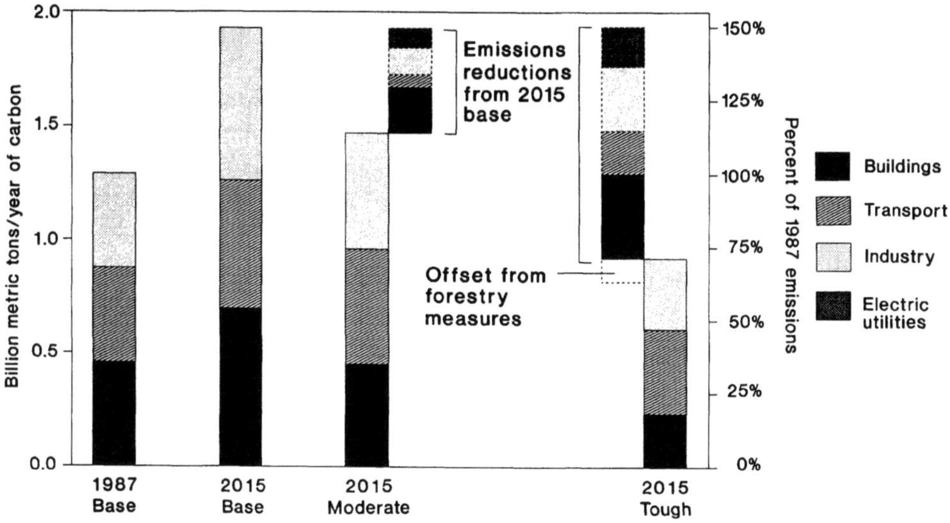

Figure 1. Carbon Emissions Under the OTA Base Case, Moderate, and Tough Scenarios. Note: The boxes outlined with dashed lines represent the *reductions* in carbon emissions associated with control measures applied in each of the three demand-side sectors (i.e., buildings, industry, and transport), and electricity supply (electric utilities); additional carbon offsets afforded by forestry measures are also shown. The boxes outlined with solid lines represent *total emissions* from each demand-side sector. Emissions associated with electricity generation (i.e., the supply-side) are included in each of the three demand-side sectors. Source: Office of Technology Assessment, 1991.

These compliance costs, when expressed as a percentage of Gross National Product (GNP), range from a savings of a few tenths of a percent to a cost of up to 1.8 percent of the GNP projected for 2015. For comparison, all environmental compliance costs today are about 1.5 percent of GNP; direct fossil fuel and electricity consumption purchases account for about nine percent of GNP.

OPTIONS FOR REDUCING U.S. EMISSIONS

The major options available or likely to be available for **reducing** CO_2 emissions in the near term fall into three categories:

1. Increasing efficiency of end-use technologies;
2. Changing use patterns to conserve energy; and
3. Shifting energy supply away from high CO_2-emitting fuels.

Additional options to **offset** CO_2 emissions are primarily forestry-related or agricultural.

When choosing policy options Congress must consider two interdependent components: the universe of possible technical changes (or in some cases, behavioral changes) and the policy instruments (e.g., taxes, regulations, financial incentives) available to require or encourage the technical change. One policy option, for example, would be to reduce CO_2 emissions through regulations (i.e., a policy instrument) to require more fuel-efficient autos (i.e., a technical change). An alternative or perhaps complementary policy option would be to use a high "gas guzzler" tax (i.e., policy instrument) to stimulate purchase of fuel-efficient autos.

Technical Options

OTA's report identifies a range of CO_2-reducing technical options available or likely to be available to the nation over the next 25 years, and what their contribution might be. There are a large number of technical options to pick from and many targets of opportunity within each sector (Figure 2). Significant progress in reducing U.S. CO_2 emissions will require that most of these options be pursued simultaneously.

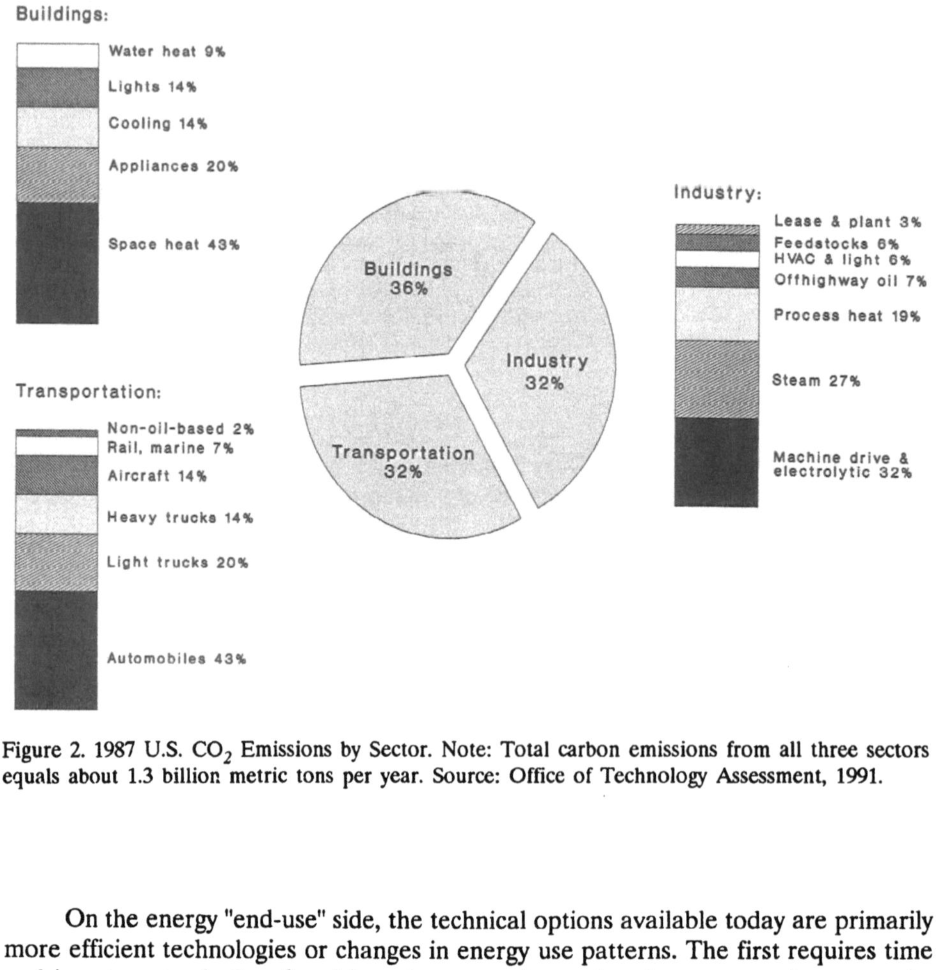

Figure 2. 1987 U.S. CO_2 Emissions by Sector. Note: Total carbon emissions from all three sectors equals about 1.3 billion metric tons per year. Source: Office of Technology Assessment, 1991.

On the energy "end-use" side, the technical options available today are primarily more efficient technologies or changes in energy use patterns. The first requires time and investment, whether for old equipment to be replaced or new equipment to be purchased. Changing energy use can include immediate (but reversible) changes such as fewer miles driven, lights dimmed, etc. In addition to the currently available technical options there is a large menu of additional options that could be developed over time.

Presently available energy "supply" options for achieving major CO_2 reductions over the 25-year timeframe of this assessment include: replacing high carbon-emitting fuels (e.g., coal) with lower carbon-emitting fuels (e.g., natural gas); using high-efficiency electricity-generating technologies (e.g., high-efficiency gas turbines or cogeneration); and using nonfossil fuels.

While nonfossil energy offers the greatest long-term potential for achieving deep cuts in CO_2 emissions, we cannot count on large-scale use of nonfossil energy sources to replace fossil fuels within 25 years. These sources do not yet offer the performance, costs or social acceptance needed to fully displace fossil fuels in such a relatively short period of time.[2] Only three nonfossil sources are presently being used on a significant scale in the United States: hydroelectric power, nuclear light-water reactors (LWRs), and biomass. Because of a combination of low baseload demand growth, cost, and environmental and social problems, no orders for new LWRs have been initiated in over a decade and there are no plans underway to build new reactors in the United States. Environmental factors set an upper limit on the number of potential new dam sites for hydroelectric facilities and on biomass production.

Several technical options are available in the forestry sector to provide some **offsets** of CO_2 emissions. Increasing forest productivity and planting new trees can result in increased carbon storage that offsets fossil-fuel related emissions. Planting short-rotation tree crops for use as biomass fuels can partially replace the use of fossil fuels in some situations. These and other forestry options have attendant uncertainties and difficulties. For example, attempts to increase productivity focus on the timber component of forests (i.e., the commercially valuable portion). However, it is unclear whether increases in timber productivity actually indicate whether or not productivity in the entire forest has increased.

Policy Instruments

Policy instruments are the means government uses to require or encourage a desired technical or behavioral response. Whatever the CO_2 reduction goal, Congress will have to use a variety of policy instruments to stimulate a diverse set of decision makers to use the appropriate fuels, technologies, and forestry and agricultural practices and to adopt energy use patterns that conserve energy. These instruments can be grouped into six generic categories: (1) taxes; (2) financial incentives; (3) marketable permits; (4) regulations; (5) research, development and demonstration (RD&D); and (6) information and public education.

Just as there is no single technical option that is sufficient, no single policy option is a cure-all. The synergisms possible among taxation, regulation, incentives, information, and RD&D programs are key to significantly reducing emissions. Taxes, if properly set, can be used to adjust prices to tilt purchase decisions. Regulation (codes and standards) can be used to remove the least efficient equipment, appliances and buildings from the market. Incentive and information programs can be used to clarify cost information and help create a market for improved energy performance. Education programs also provide consumers with the knowledge and information needed to make wiser energy choices. Government-sponsored RD&D can help provide producers and consumers with new technical options that can be used to reach national goals, as well as reduce, by cost-sharing, the risk to industry of developing these new options.

Three of these instruments are applicable to all sectors: (1) carbon taxes; (2) marketable permits for carbon; and (3) government-funded research and development for noncarbon energy sources and energy efficient technologies. Sector-specific options are discussed later in this paper.

Carbon Taxes

Carbon taxes offer a way to make high CO_2-emitting technical options more expensive than lower CO_2-emitting options. The tax is levied on the carbon contained in fossil fuels and is set to reflect the fact that some fuels emit more carbon per unit of energy than do others. The highest taxes would be on coal. The tax on petroleum would be about three-quarters of the tax on coal and the tax on natural gas would be about half that of coal.

Carbon taxes have several advantages. First, they are a relatively simple way of affecting all emitters simultaneously, rather than focusing on a few selected technologies. A carbon tax also targets the heaviest economic sanctions against the worst emitters of CO_2. Carbon taxes can be used alone or in combination with other sectoral-specific approaches listed below. If Congress so desires, new tax monies could help fund incentive programs, offset the budget deficit, or replace other existing taxes.

Marketable Permits

Marketable permits and carbon taxes are closely related. Under a marketable permit system, policymakers fix the total amount of carbon that can be emitted. The government then issues a number of carbon emission permits equal to that limit. Permits can be bought and sold by energy users just like fuels. For example, for every 1 million Btu of coal purchased, the user must also own (or purchase) permits to emit 57 pounds of carbon. To burn 1 million Btu of natural gas, the user must own or purchase permits to emit 32 pounds of carbon. If demand for energy rises, the price of a carbon permit will rise to reflect the cost of lowering emissions. Some holders of permits will find ways to lower emissions (e.g., purchase more efficient equipment, switch from coal to natural gas) so that they can sell their permits (at a profit) to others.

A marketable permit system is the regulatory mechanism for limiting emissions of sulfur dioxide to control acid rain under the new Clean Air Act Amendments (Public Law 101-549).

Research, Development, and Demonstration (RD&D)

Through RD&D, government can search for and fine-tune technological fixes to the greenhouse gas emissions problem. In fact, climate change can only be effectively addressed over the long-term with the development and worldwide use of better nonfossil energy sources. Government can speed the process of testing and commercializing many energy-supply and end-use technologies. However, only about five percent of the $2.7 billion national budget for energy technology R&D in 1990 was devoted to renewables (including biomass energy) and only seven percent to energy conservation. Fossil fuels had 25 percent of the research budget, nuclear fusion 12 percent, and nuclear fission 9 percent.

Table 1. Measures to Lower U.S. Carbon Emissions (expressed as percentage of 1987 total emissions)[a]

	Reductions in 2015	
	Moderate (in percent)	Tough (in percent)
DEMAND-SIDE MEASURES		
Residential buildings		
New investments:		
Shell efficiency	1.3	2.0
Heating and cooling equipment	0.1	0.4 to 0.6
Water heaters and appliances	1.2	1.5 to 2.3
O&M, retrofits:		
Shell efficiency	0.8	0.9
Lights	0.6	0.8
All residential measures together	4	5.6 to 6.6
Commercial buildings		
New investments:		
Shell efficiency	2.3	4.0
Heating and cooling equipment	1.0	1.2 to 1.9
Lights	2.1	3.0
Office equipment	1.6	2.1
Water heaters and appliances	0.1	0.1
Cogeneration	0.2	1.5 to 2.3
O&M, retrofits:		
Shell efficiency	0.8	0.8
Lights	0.5	0.5
All commercial measures together	8.5	13 to 15
Transportation		
New investments:		
New auto efficiency	0.8	3.5 to 3.8
New light truck efficiency	0.5	2.5 to 2.7
New heavy truck efficiency	0.4	2.4 to 2.4
Non-highway efficiency	0.5	1.2
O&M, retrofits:		
Improved public transit	0.2	3.5
Truck inspection & maintenance	0.3	0.4
Traffic flow improvements/55 mph speed limit	1.2	1.4
Ridesharing/parking controls	0.4	1.0
All transportation measures together	4	14 to 15
Industry		
New investments:		
Efficient motors	1.2	3.7 to 4.0
Lighting	0.5	0.7 to 0.8
Process change, top 4 industries	3.0	8.2
Fuel switch to gas	0.0	2.4 to 2.7
Cogeneration	0.8	5.2 to 5.8
O&M, retrofits:		
Housekeeping	1.9	2.0
Lighting	0.1	0.2
All industrial measures together	8	17 to 18
ELECTRIC UTILITY SUPPLY-SIDE MEASURES		
Existing plant measures:		
Improved nuclear utilization	4.1	4.1
Fossil efficiency improvements	1.7	1.7
Upgraded hydroelectric plants	0.5	0.5
Natural gas co-firing	—	3.7
New plant measures:		
No new coal; higher fraction of new nonfossil sources	—	0.0 to 4.7
CO_2 emission rate standards	0.4	0.0 to 0.1
All utility supply-side measures together	6.6	9.9 to 14
FORESTRY MEASURES		
Afforestation:		
Conservation Reserve Program	0.2	0.2
Urban trees	—	0.7
Additional tree planting	—	2.3
Increased timber productivity	—	3.1
Increased use of biomass fuels	—	1.2
All forestry measures together	0.2	7.5

[a] 1 percent of 1987 emissions = 13 million metric tons C = 0.7 percent of 2015 emissions.

SOURCE: Office of Technology Assessment, 1991.

Sectoral Policy Options

Buildings Sector Policy Options

Table 1 summarizes the emissions reductions possible for each technical option modeled by OTA under both the moderate and tough scenarios. For buildings, improving shell efficiency and lighting are the two technical options with the greatest potential for lowering CO_2 emissions. Under the base case, OTA assumes that by 2015 new homes and apartments will be designed such that they need about 15 percent less heating and 8 percent less cooling than current new homes. By adopting moderate shell efficiency measures, such as thicker insulation and better windows, new homes will require an estimated 50 percent less heat and 25 percent less air-conditioning than today's average new home.[3] With tough measures, homes can be built to require an estimated 85 percent less heat and 45 percent less air-conditioning.[4]

As shown in Table 1, moderate shell improvements in new residential buildings can reduce U.S. carbon emissions by 1.3 percent of current CO_2 levels by 2015. By implementing tough improvements in the North and moderate ones in the South, reductions of two percent in new residential buildings might be achieved. Tough measures for new commercial building shells can achieve reductions equal to four percent of 1987 levels by 2015.

Existing homes can also be made more efficient by installing more efficient heating and cooling equipment, insulation, windows, etc. The base case assumes that existing homes will require six percent less heating by 2015 because of replacements and improvements that will happen anyway. Moderate measures boost this to 25 percent by 2015 and tough measures boost it to 40 percent by 2015.[4] Tough measures in the North and moderate ones in the South would reduce carbon emissions from existing buildings by four percent by 2000, but would have a declining effect thereafter as many of the older homes are replaced by new ones.

Improving the efficiency of lighting in new commercial buildings is another technical option that can yield substantial reductions. The tough scenario measures together—a combination of high-efficiency fluorescent bulbs and ballasts, improved reflectors, and better use of daylight—would lower lighting energy needs by 60 percent in these buildings.[5] This achieves reductions equal to three percent of 1987 emissions by 2015.

Gains in buildings can also be made by simply replacing existing bulbs with high-efficiency ones—without replacing fixtures—as shown under the "Operation and Maintenance" heading of Table 1. Replacing the most heavily used incandescent bulbs in homes with compact fluorescents and using high-efficiency fluorescents in commercial buildings can lower emissions by 1.3 percent under our tough scenario.

The policies described below appear to offer the most promise to achieve these reductions. While a general carbon tax will certainly help, because there are so many different decision makers—some of whom may not be that responsive to price changes—a larger arsenal of policy instruments is needed. These include demand-side management (with the utilities as partners) as well as a series of targeted financial sanctions, incentives, and regulations.

Demand-Side Management (DSM)

DSM refers to electric utility programs designed to encourage customers to modify their patterns of energy use. Particularly promising—from a global warming perspective—are those situations where utilities allow energy conservation to compete with traditional supply technologies (e.g., power plants) to balance energy supply and demand. DSM can be an effective approach to reduce energy consumption by improving building shells as well as the equipment inside buildings. In some cases, utilities pay for rebate programs, give out high-efficiency light bulbs, or otherwise stimulate end-use efficiency improvements, and save energy at a fraction of the cost of new power supplies.

Demand-side management can result in greater investments in energy efficiency than customers would otherwise make. Utility programs have long time horizons and can capture the potential in both the new and retrofit markets, for both equipment efficiency and building shell improvements. There is already considerable support for DSM by many State energy offices, State legislatures, and public utility commissions.

However, in order for DSM to stimulate significant investment in conservation, incentive structures must be changed so that utilities can profit from demand-side investments. Any federal legislation concerning DSM would need to be general enough to allow States flexibility in implementation and specific enough to have a genuine impact on conservation. Congress could provide funding to evaluate various incentive structures currently being examined by States and utilities. Should Congress wish to pursue more direct action, it could require States to formally consider demand-side resources in their planning, with oversight by the Federal Energy Regulatory Commission (FERC).

Further, the federal government could mandate that environmental externalities be considered in evaluating supply-side options (as New York State has done—i.e., penalizing polluting options based on estimates of the costs of environmental damage that would accrue).[6] Congress has already mandated, in the 1980 Pacific Northwest Electric Power Planning and Conservation Act (Public Law 96-501), that the Northwest Power Planning Council adopt rate structures that give conservation measures a cost break over other, more traditional supply-side measures.

Technology-Specific Regulations

Congress can directly mandate efficiency improvements through appliance standards and building energy codes.

Appliance Standards. Appliance standards, by fiat, remove inefficient appliances from the market. The National Appliance Energy Conservation Act (Public Law 100-12), passed in 1987, is expected to lower residential energy use by up to ten percent by the year 2000.[7] However, even stricter standards are possible. The law requires review of appliance standards twice during the 1990s, which provides an opportunity to obtain additional energy reductions through more stringent standards. Congress could also consider extending standards to other equipment such as commercial heating, ventilation, and air-conditioning equipment; light bulbs; and building components such as windows.

Standards could be even more effective if used in conjunction with other incentives. Policies such as utility programs, appliance labeling, and tax schemes provide incentives to do more than standards require.

Building Energy Codes. Building energy codes serve a function analogous to that of appliance standards by preventing the least efficient buildings from being constructed. Building codes have traditionally been under the jurisdiction of states and localities. Currently, there is little support from the states or the construction industry for a mandatory national building code. In 1976, Congress enacted legislation that required the development of the Building Energy Performance Standards (BEPS), a mandatory national code based on performance standards. In 1983 the law was modified to be a mandatory code only for federal buildings.

Financial Measures

Congress can choose from among several sector-specific financial mechanisms, including building tax credits and subsidies and initial purchase taxes for appliances and other equipment.

Building Tax Credits and Subsidies. Tax credits and subsidies for using more efficient technologies can promote retrofitting of existing residential and commercial buildings. The federal government, for example, passed legislation that provided solar and conservation tax credits for the years 1978 through 1984. By 1983, 24 million households claimed a residential tax credit of up to $700 each for investments in energy conservation; however, no evaluation or monitoring of energy saved by this program was ever conducted.

Initial Purchase Taxes and Rebates for Appliances and Other Equipment. An initial purchase tax scaled to penalize inefficient equipment could accelerate the market penetration of efficient equipment. Examples include a lump-sum tax on appliances and equipment at the time of purchase. Taxes collected on the most polluting items could be used to provide a rebate on the least polluting items. However, although an initial purchase tax sends appropriate signals regarding consumer purchasing decisions, it would not—unlike an energy or carbon tax—change use of an appliance once it is purchased.

Consumer Information and Marketing Programs

Lack of information and uncertainty have been identified as key barriers to greater investment in energy conservation in the buildings sector. The large number of highly cost-effective investments in energy efficiency that are not chosen by consumers indicates that price alone does not stimulate optimal investment decisions. Therefore, information dissemination is a key element of several of the policy options discussed above, including the sector-specific financial measures and general energy and carbon taxes.

Home Energy Rating Systems. The federal government has been involved in home energy rating systems—which tell buyers how efficient their prospective homes or offices are—through its role in the mortgage market. In addition, the National Affordable Housing Act of 1990 requires HUD to develop a plan to make housing more affordable through mortgage financing incentives for energy efficiency. The federal government could play a further role by developing a uniform energy rating system for all residential and commercial buildings, making it easier and less expensive for lenders to include energy costs in their mortgage evaluations.

Energy Audits. The Federal Residential Conservation Service, created in 1978, mandated that gas and electric utilities provide their customers with on-site energy audits. The program was implemented in 1981 and recently expired. There has been little evaluation of the program, and little reliable information has been kept on its success in reducing energy consumption. However, while it is unclear whether information from audits alone is enough to encourage conservation, it would certainly seem to be useful when combined with other measures.

Building Research, Development, and Demonstration

Major barriers to private investment in RD&D in the buildings sector include the fragmented structure of this sector and the short-term perspective of many of the decision makers (e.g., builders, renters). In addition, the U.S. government currently spends a negligible amount on housing research. In contrast, in countries such as Sweden and Japan, RD&D spending has been part of a trend toward energy-efficient prefabricated housing. This spending has contributed to the energy efficiency of homes through standardization of energy-saving features and quality control in the design and manufacture of building components.

The Federal Energy Management Program (FEMP), administered by the Department of Energy, works with government agencies to implement cost-effective, energy-efficiency improvements. Congress could authorize FEMP to test and demonstrate performance, acceptance, and cost-effectiveness of new technologies in federal buildings.

Transportation Sector Policy Options

Urban passenger travel in cars and light trucks (i.e., light vehicles) in the United States requires the largest share of transport energy, consuming 15 percent of the world's oil production. The two main opportunities for reducing transportation's contribution to global warming are measures to increase the energy efficiency of light vehicles and measures to encourage urban passengers to drive less. Thus, under OTA's modeling exercise, the major reductions in this sector come from higher auto and truck efficiency, better control of traffic, and, under the tough scenario, more use of public transit (see Table 1).

With respect to auto efficiency, our base case assumes that new cars will average about 32 mpg by 2000 and about 37 mpg by 2010. Under the moderate scenario, new car efficiency averages 35 mpg by 2000[8] and 39 mpg by 2010.[9] Under the tough scenario, we assume a range of new car efficiencies. For example, efficiencies of 39 mpg by 2000 and 55 mpg by 2010 might be possible even if consumers maintain their current preferences for car size and performance.[9] If consumers are willing to buy smaller cars, new car fleet average efficiencies of 42 mpg by 2000 and 58 mpg by 2010 might be achievable. Given this range of assumptions, reductions amount to about 3.5 to 3.8 percent of current CO_2 emissions by 2015 (see Table 1). Reductions of about 2.5 to 2.7 percent from light trucks and another 2.4 percent from medium- and heavy-duty trucks are achievable under our tough scenario, as well.

Traffic speed affects fuel consumption, too. By reinstating the 55 mile-per-hour speed limit and by reducing traffic congestion in urban areas in order to speed up travel, reductions of 1.4 percent by 2015 are possible under our tough scenario.

Measures to move people out of their cars and into mass transit under the tough scenario would yield reductions of about 3.5 percent by 2015. To achieve this, however, urban auto traffic would have to be reduced by ten percent through urban light rail and improved urban design. Additionally, five percent of car travel between cities would have to shift to high-speed intercity rail.

The following four policy instruments will promote new car efficiency: gasoline taxes, vehicle taxes and rebates, fuel economy standards, and incentives for vehicle manufacturers. In addition, improved operation and maintenance practices will reduce energy use in existing cars. Two other measures, transportation control measures (TCMs) and controlling settlement patterns, can help reduce CO_2 by reducing vehicle miles traveled.

Gasoline Taxes

A gasoline tax would create incentives for both increased efficiency and reduced travel. Taxes would induce consumers to use less fuel while leaving them free to choose how they adjust their behavior. In concert with increasing fuel economy standards (see below), a long-term impact on the efficiency of the vehicle fleet could be achieved.

Although the effectiveness of taxes is hard to predict from studies of past responses to price changes, one might expect a ten-percent hike in gasoline prices to yield a one- to six-percent drop in gas consumption.[10,11] A 50-percent increase in price might reduce consumption 5 to 20 percent over the near term, even more over the longer term. A doubling or tripling in price (approaching the costs in Europe and Japan) might yield an immediate decrease of 13 to 20 percent and a longer term response of a 35- to 40-percent reduction in gasoline consumption. About half of this longer-term adjustment to high price is expected from driving less, and the other half from more efficient vehicles. For example, consumers might choose to spend money on fuel-efficient technologies or to use mass transit, carpool, or simply travel less.

Vehicle Taxes and Rebates

Taxes on inefficient vehicles can create incentives to choose better fuel economy and forego large size and extra power. Such a program would be most effective if accompanied by rebates for highly efficient cars. In a "revenue neutral" program, the money taken in from the taxes would be redistributed through the rebates. The Federal Gas Guzzler Tax already applies to cars with fuel economies below certain thresholds; the Omnibus Budget Reconciliation Act of 1990 recently doubled the tax for cars getting less than 22.5 mpg.

An expanded program of auto purchase taxes and rebates could complement fuel economy standards and taxes, but it could also pose serious trade difficulties as long as the high-efficiency end of the auto market is dominated by imports. If implemented suddenly, such measures would put domestic manufacturers at a disadvantage; on the other hand, measures designed to protect domestic manufacturers might conflict with General Agreement on Tariffs and Trade (GATT) rules.

Fuel Economy Standards

Standards influence the tradeoffs among cost, performance, size and efficiency that underlie the development and introduction of new models. The current fuel economy standards for cars, in place since 1978, have helped to increase auto fuel economy.[12] More stringent standards can both lower CO_2 emissions and reduce our dependence on imported oil. Redesigned standards that vary with vehicle volume can help minimize the burden on U.S. manufacturers that offer a full range of car sizes.[13,14]

Incentives for Vehicle Manufacturers

One incentive, aimed at producers instead of consumers, is the use of government-sponsored competitions to induce manufacturers to develop high-efficiency or alternate-fueled cars. A variant of the incentive scheme injects competitive elements into a high-efficiency rebate program. For example, the government could identify a few classes of vehicles most in need of fuel economy improvement and offer a competitive reward in the form of consumer rebates on a large (e.g., 200,000 units) production run of a new vehicle achieving the best fuel economy above a specified threshold.

Efficient Vehicle Operating Practices

Changes in vehicle operating practices offer small potential reductions individually but often have short startup times and do not require large, up-front capital investment. They include reimposing (and enforcing) the 55-mph speed limit, requiring efficiency inspections for trucks, and charging efficiency-promoting parking fees at federal offices.

Transportation Control Measures (TCMs)

TCMs include a wide variety of measures to reduce the number of vehicle miles traveled (VMT) and lower congestion. They are attractive because they typically have short startup times and low capital costs, and can reduce energy use and greenhouse emissions even within existing settlement and employment patterns. In aggregate, TCMs appear to hold modest promise for reducing VMT. These include such measures as: ridesharing (promotion and matching services); employer-based transportation management (high parking charges or transit or vanpool subsidies); High-Occupancy Vehicle (HOV) lanes (restricting lanes on freeways to cars with three or four occupants and to buses); park-and-ride (intercept drivers near their origins); and mass transit improvements (bus service expansion, operational changes, and fare changes).

Controlling Settlement Patterns

Long-term reductions in emissions can be achieved by changing patterns of settlement to reduce the need for travel. This can be accomplished through higher densities, or through mixing uses so that residences, jobs, and services are roughly balanced at a local scale. When more destinations are close to home, more trips can be made by foot; when densities are higher, public transit can serve more people effectively.

In the United States, except possibly for some high-growth areas in the South and West, efforts to change the shape of settlement in major cities may meet local resistance. Nevertheless, some changes are feasible in suburban areas nationwide. Stringent suburban restrictions on development—sometimes only on commercial and industrial development, sometimes on new residential development as well–have been attempted in some regions of the United States.[15]

Transportation RD&D

American automakers lag behind their Japanese, and to a lesser extent their European, counterparts in moving research results to the market.[16] In the 1980s, a program to support more aggressive research and development in the American auto industry—the Cooperative Automotive Research Program—was briefly attempted by the Department of Transportation. A revived, combined government/industry program could be successful if domestic automakers, their suppliers, and innovative research companies all are key players. The program could target important efficiency areas such as continuously variable transmissions, energy-storage systems, new engine designs for heavy trucks, improved safety for lighter vehicles, and innovations to permit increased intermodal freight.

An area of longer-term research that deserves special attention is development of truly clean, economically acceptable, alternative fuels and a supporting infrastructure. Fuels with the greatest potential—electricity or hydrogen from noncarbon energy sources (e.g., solar and nuclear) and woody biomass fuels grown on a sustainable basis—are the furthest from large-scale technical viability. Research in these areas could be expanded, with parallel programs to assess and demonstrate the actual performance of a variety of fuels.

Manufacturing Sector Policy Options

For manufacturing, as shown in Table 1, three types of technical improvements offer the greatest promise. The first area is "process changes." The top four manufacturing energy consumers (paper, chemicals, petroleum and primary metals)—which account for more than 75 percent of energy use in this sector—improved their energy efficiency by between 2.3 and 4.3 percent per year between 1980 and 1985.[17] If this pace can be maintained, as we assume in our tough scenario, reductions equal to about eight percent of current emissions (by 2015) will result.

Cogenerating electricity and steam for industrial processes is a second promising option. When utilities generate electricity, about two-thirds of the energy from burning the fuel is released as heat. If electricity is generated at industrial sites where the heat can be used, the efficiency of fossil fuel use can be increased dramatically. Under our tough scenario, we assume that 90 percent of new and replacement industrial steam boilers will cogenerate electricity. Such measures can lead to reductions equivalent to about 5.8 percent of current total U.S. emissions.

More efficient motors are a third technical improvement that can lead to substantial improvements. Moderate and tough measures might improve motor efficiencies by 10 percent and 30 percent,[18] respectively, yielding reductions of about 1.2 percent by 2015 under the moderate scenario and 4 percent under the tough one.

The following policy instruments could encourage these technical measures: carbon taxes, DSM, efficiency standards, marketable permits, tax incentives, informational policies, and RD&D.

Carbon Tax

A carbon tax would levy economic penalties against the highest industrial emitters of CO_2. Under such an approach, the tax would be highest on coal, low for natural gas, and zero for noncarbon sources (e.g., wind, solar, geothermal, or nuclear). For industries where the cost of energy is particularly important, carbon taxes should encourage energy efficiency, fuel switching and cogeneration.

Using several econometric models, the Congressional Budget Office estimated that a carbon tax of $100 per ton would lower CO_2 emissions from industry by between 10 and 35 percent by the year 2000. The higher reduction estimate reflects a 70-percent reduction in coal use.

Demand-Side Management

DSM programs—joint programs between electric utilities and their customers discussed previously—can help lower electricity use in the industrial sector. The major programs include: (1) rebates to customers who install approved equipment; (2) low-interest loans to customers for conservation installations; and (3) installation of conservation equipment at utility cost.[19]

Standards

A more traditional regulatory policy is to require efficiency standards for common energy-using equipment, similar to those existing for automobiles and some appliances. Motors would be the most likely candidate for this approach.

Marketable Permits

CO_2 emissions can be regulated by requiring permits for emissions; manufacturers could be issued permits based, for example, on some percentage of their 1990 emissions. Reductions might be accomplished by installing energy-efficient technologies and fuel switching; offsets could result from approved reforestation/afforestation projects. It would be up to the manufacturer to choose the most cost-effective strategy. Marketable permits would allow firms to trade their unused carbon rights to a firm that is exceeding its budget.

Manufacturing Tax Incentives

Much industrial equipment is old and energy-inefficient compared to the best available technology. In many cases, replacing old equipment improves energy efficiency by 10 to 50 percent. Financial policies, such as tax credits or accelerated tax depreciation schedules, aimed at stimulating rapid replacement of older equipment have the potential to stimulate improvements in energy use. Such policies have a precedent: the Energy Tax Act of 1978 provided a ten-percent added "energy investment tax credit" for certain energy-conservation investments (as well as tax credits for some energy-supply investments). The tax credits were available until 1985 and applied to a specific list of technologies. However, rather than specify which technologies qualify, Congress could foster innovation by offering similar—or greater—tax breaks for company-chosen conservation technologies.

Informational Policies

A barrier to reducing emissions in the manufacturing sector is lack of information about how to improve energy use—especially for smaller, less energy-intensive industries. Informational policies can include performance goals, the collection of performance data, labeling of the energy performance of equipment, training, and performance audits.

Renewed support for cooperative government/industry information-sharing programs could help. For example, DOE's Energy Analysis and Diagnostic Center program funds faculty and students at several universities to perform free energy audits for small and medium-sized manufacturers in more than 30 states. Because costs saved by manufacturers translate to increased taxable income, the program can provide additional tax revenues to the federal government. The biggest cost savings have come from efficiency improvements associated with cogeneration, space heating, lighting, and process equipment maintenance and replacement (in descending order of savings).[20] This program could be expanded or new programs could be modeled after it.

Manufacturing R&D

Research and development sponsored by DOE's Office of Industrial Programs in waste energy reduction and industrial process efficiency, if funded, are projected to save more than three to four percent of energy used by industry per year over the next decade. Research areas identified by Oak Ridge National Laboratory as particularly promising are: improved use of catalysts in chemical production; intelligent sensors and controls; and heat recovery and cogeneration.[21] R&D in non-energy areas, such as materials science, also holds promise for partial replacement of energy-intensive materials like steel and aluminum. Likewise, research and development to improve the quality of products made with recycled materials could help reduce energy use by increasing the demand for recycled materials such as paper, steel and aluminum.

Electricity Generation

About one-third of U.S. carbon emissions come from generating electricity; by 2015 under our base case this may be as high as 45 percent. Thus measures that lower the rate of carbon emissions per kilowatt-hour (kWh) of electricity generated would translate into substantial reductions.

Table 1 shows OTA's estimate of the technical potential for emissions reductions in the electric utility sector, depending on the demand for electricity and the stringency of policies. Moderate utility supply-side measures can lower emissions by about 6.6 percent. The two with the greatest reduction potential are: (1) increasing the efficiency of fossil-fuel-fired plants (by about five percent) through improved maintenance;[22] and (2) operating existing nuclear power plants 70 percent of the time, similar to Western Europe and Japan,[21] and extending their useful life to 45 years.

Our tough measures eliminate coal use wherever possible. A combination of renewable energy sources, nuclear plants with improved designs that may be available after 2005, and high-efficiency gas turbines are the only new utility plants built under the tough scenario. However, if all the tough demand-side measures in the buildings and industrial sectors are implemented, growth in demand for electricity is so low that very few new plants are needed through 2015. Thus, the only way to lower emissions under this scenario is to either co-fire existing coal plants (e.g., with 50 percent natural gas), or retire existing coal plants after 40 years of operation (rather than the typical 60 years) and replace them with renewable or nuclear fuels or natural gas. The former measure would reduce emissions by about 3.7 percent by 2015; the latter, by about 4.7 percent of current levels by 2015.

The following policy options could be used to encourage these technical measures: carbon taxes, marketable permits, subsidizing noncarbon sources, emissions limits and standards, and RD&D.

Carbon Taxes

A carbon tax, if set high enough, would encourage fuel switching and conservation. A carbon tax in the range of $75 to $150 per ton would make natural gas a more economic choice than coal at many facilities. A carbon tax would also provide added motivation to develop more noncarbon energy sources.

Marketable Permits

Utilities could be issued marketable permits for CO_2 emissions allowed from their coal-fired units, based on their generation in a historic year (e.g., 1990) multiplied by an allowed emission rate. Under this approach some utilities could curtail coal use more than necessary to meet their limits and sell permits to others exceeding their limits.

Subsidize Noncarbon Sources

Any of the general financial instruments, such as a carbon tax or fossil fuel energy tax, will serve to encourage use of nonfossil sources for electricity generation. According to one estimate,[2] a 2 cent-per-kWh subsidy or its equivalent for only renewable sources of electricity might double the contribution of renewable sources of electricity by 2010—i.e., allow them to supply 40 percent of new demand under a base case growth scenario. Under our tough scenario, we assume nonfossil sources can provide between 30 and 45 percent of new demand (depending on the success of demand-side measures).

CO_2 Emission Limits and Efficiency Standards

Congress could mandate reductions by setting CO_2 emission limits or efficiency standards. For example, an emission rate limit of 0.55 pounds carbon per kilowatt-hour (lbs C/kWh) would require a typical midwestern plant burning Illinois coal to burn between about 10 and 30 percent gas, depending on its efficiency. At 0.55 lbs C/kWh, the most efficient new coal burning technologies (e.g., integrated coal gasification combined cycle, or IGCC) would just qualify burning coal alone.

Two somewhat different strategies could be pursued to set CO_2 emission limits for new plants. If the intent is to force development of ultra-efficient coal technologies, then a standard in the range of 0.35 to 0.40 lbs C/kWh would be appropriate. Molten carbonate fuel cells, if successful, might be able to achieve such emission rates using bituminous coals. If the intent is to limit new fossil-fuel-fired generation to the cleanest sources only—advanced combined cycle turbines burning gas—then a new source performance standard of about 0.25 lbs C/kWh would be more appropriate. To speed up replacement of old plants with new, lower emitting ones, Congress could mandate the retirement of existing fossil-fuel-fired plants earlier than their expected lifetime of sixty years.

Energy RD&D Funding

Over the last decade, federal funding for renewable energy, conservation and nuclear (fission) R&D fell rapidly. The 1990 combined energy technology R&D budgets (in 1990 dollars) for these three categories were 82 percent lower than they were in 1980 (see Figure 3). To reinstate the funding levels of ten years ago would require adding about $2.6 billion. By doing so, the federal government could hasten the development and demonstration of supply technologies that would reduce greenhouse gas emissions. The most promising of these technologies include: commercial fuel cells; storage technologies for solar electricity; biomass-driven turbines; variable-speed wind turbines; and better designs for nuclear power plants. Many experts estimate that these technologies could be commercially available within the next few decades.

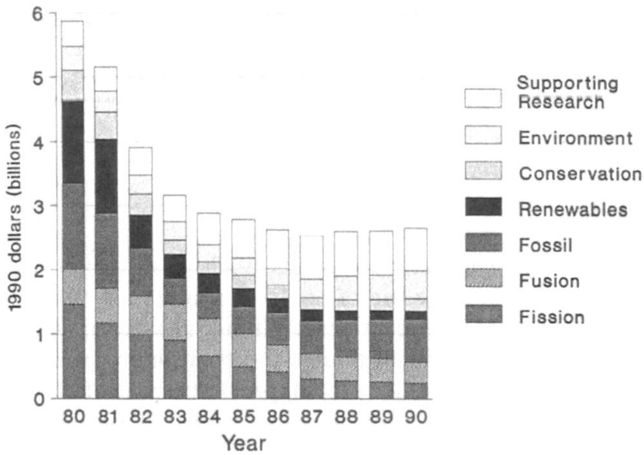

Figure 3. Federal Government Funding for Energy Supply R&D, 1980 to 1990. Note: The bars represent federal budget authority for research, development, and technology demonstrations, in 1990 dollars. "Supporting" refers to research in basic energy sciences. Sources: J.P. Holdren, "Energy in Transition," *Scientific American* 263(3):156-163, September 1990, original data compiled from the Office of Management and Budget, *Budget of the United States Government, Fiscal Years 1980 Through 1991* [separate volume for each year] (Washington, DC: U.S. Government Printing Office, 1980 through 1990); Intersociety Working Group/American Association for the Advancement of Science, *Research and Development FY 1980-1991* [separate volume for each year] (Washington, DC: 1980-1990); and *Environmental and Energy Study Conference, Weekly Bulletin(s)* and *Special Report(s)* [weekly and occasional, respectively] (Washington, DC: several years).

The government could also play a role in reducing the perceived risk of new technologies and integrating renewable energy sources in existing energy systems by conducting demonstration projects or perhaps providing government-backed loans. To encourage new nuclear energy sources, a two-track process appears best: the Department of Energy could help fund full-scale demonstrations of both new "evolutionary" light-water reactors and "revolutionary" design changes such as a modular high-temperature gas reactor.

For existing nuclear power plants, the goal should be to increase the number of hours of operation, rather than to increase efficiency of fuel use. A Department of Energy demonstration program (coordinated with the Nuclear Regulatory

Commission) might bring U.S. hours of nuclear plant operation from well below to above the average for Western Europe and Japan. Key elements of such a program would include improving preventive maintenance, installing automated controls to improve reactor operation, and speeding up time spent refueling.

Forests

Forestry-related measures with the greatest potential to offset carbon emissions include increasing the productivity of existing forests, planting trees in new areas, and growing tree crops for biomass energy; we consider these to be tough measures, with the exception of ongoing tree planting in the Conservation Reserve Program. OTA estimates that the increased carbon uptake from increasing productivity on about 60 million hectares of timberland might be equivalent to annual emissions reductions of about three percent of current levels by 2015. Planting new trees (i.e., afforestation) on farmland, on nonforested areas and in cities might result in carbon storage equivalent to emissions reductions of about two percent of current levels by 2015. Planting trees for biomass energy might result in an additional reduction of about one percent by 2015.

There are several caveats to this potential for offsetting emissions. Trees planted today can continue to store carbon beyond this study's 25-year timeframe. But this carbon eventually will be released to the atmosphere, either when trees die and decompose naturally, when they are harvested and burned, or when products made from wood eventually decompose. Unless the wood is used to displace fossil fuel use or is permanently stored under conditions that do not allow decomposition, carbon offsets in later years will dwindle. These estimates also assume that increasing the productivity of a forest's commercial timber component is equivalent to increasing the productivity of the entire forest ecosystem, but this assumption needs to be tested. Finally, forests—and the feasibility of using forestry practices to offset emissions—are likely to be affected by future climate changes. Therefore, forestry options in industrialized countries such as the United States cannot be considered a substitute for reducing total energy use, but rather as a way of "buying" time while developing alternative nonfossil fuel sources and improving the efficiency of energy use in general.

Congress could promote management practices that increase carbon storage or offset CO_2 emissions by augmenting existing forest management and tree planting programs of the U.S. Forest Service (USFS) and the Agricultural Stabilization and Conservation Service, and by enhancing the biomass energy research program of the Department of Energy. In addition to direct support for such programs, Congress also could consider using financial mechanisms (e.g., changing income tax policies to encourage more investments in forest management; imposing a tax on fossil fuels to make biomass fuels more competitive).

Incentives to Increase Carbon Storage on Forest Lands

Incentives to increase productivity—i.e., net carbon storage—will differ for publicly and privately owned forests. On public lands, which are located mostly in the West, management objectives are determined by planning processes legislated by Congress. Government investments in these lands are likely to focus on reforestation and timber stand management. Congress could direct the USFS and Bureau of Land Management to, for example, increase reforestation activities and to conduct research on the ability of "new forestry" practices that proponents contend might help to both maintain higher levels of diversity and allow commodity production.

Privately owned forests are most extensive in the East and South. For nonindustry private forests, Congress could continue to increase assistance to states and private landowners under programs such as the Forestry Incentives Program and the Agricultural Conservation Program. These programs currently reach only about two percent of nonindustry private owners,[23] even though these owners undertake over 40 percent of all reforestation. The Interior and Related Agencies Appropriations Act for fiscal year 1991 almost doubled funding for the USFS's state and private forestry programs, which include tree planting and management. The 1990 Food, Agriculture, Conservation and Trade Act (Public Law 101-624), known as the 1990 Farm Bill, also authorized a forestry stewardship program in which the USFS would work with state and local governments, land grant universities, and the private sector to improve resource management on privately owned forest land.

For industry-owned timberland, investments might be stimulated through changes in capital gains provisions (e.g., restoring preferential tax rates or providing a partial exclusion from taxable income for timber held longer than twenty years) or allowing full annual deductions for expenses, as well as by increasing funding for federal assistance programs. One possibility for increasing support of such programs is to use funds that would accrue if below-cost timber sales in National Forests were eliminated.

Incentives for Growing New Trees on Unforested and Urban Lands

Mechanisms to promote afforestation include the Conservation Reserve Program (CRP), the President's proposed America the Beautiful program, and financial incentives such as tax credits for carbon storage. In general, any tree-planting program needs to consider the costs of maintaining trees in a healthy state once planted; this will be even more critical as climate changes occur.

Incentives for Biomass Energy to Offset CO_2 Emissions

Growing short-rotation woody crops on nonforested land for use as an energy source shows some promise. Congress could increase funding for Department of Energy research on uncertainties regarding long-term productivity, including effects on nutrient availability, and costs. Increasing fossil fuel taxes would make biomass fuels more competitive. Even then, farmers wishing to invest in biomass crops may be limited by loss of base acreage in commodity support programs and by lack of revenues for several years. Thus, changes in support programs or provision of some subsidy may be needed to stimulate investments in biomass crops on current cropland.

COMPARISON TO OTHER STUDIES

In addition to the OTA study, two other recent efforts looked at the interactions of energy policy and CO_2 emissions: the *National Energy Strategy*[24] and a recent report from the National Academy, *Policy Implications of Greenhouse Warming*.[25] All three studies conclude that the United States can significantly slow the growth of CO_2 emissions over the next 25 years by adopting a series of low-cost measures.

Again as shown in Figure 4, under the base case, CO_2 emissions are forecast to grow by about half by 2015. Under the fully implemented National Energy Strategy, CO_2 emissions increase 30 percent by 2015. OTA's moderate scenario and OTA's estimates of the Academy's low-cost options would hold CO_2 emissions to about 10 percent above 1990 levels. Finally, OTA's tough scenario results in a 35 percent decrease below 1990 levels.

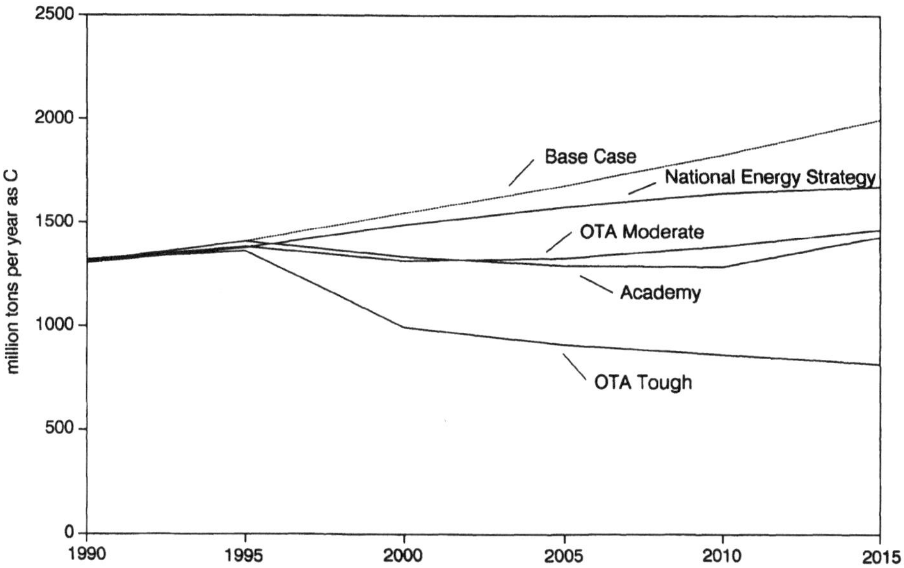

Figure 4. Comparison of Carbon Emissions under the National Energy Strategy, the National Academies' scenario, and the OTA scenarios. Sources: U.S. Department of Energy, *National Energy Strategy*, First Edition, 1991/1992 (Washington, DC: U.S. Government Printing Office, 1991); Policy Implications of Greenhouse Warming--Synthesis Panel, Committee on Science, Engineering, and Public Policy, *Policy Implications of Greenhouse Warming* (Washington, DC: National Academy Press, 1991); U.S. Congress, Office of Technology Assessment, *Changing By Degrees: Steps to Reduce Greenhouse Gases*, OTA-0-482 (Washington, DC: U.S. Government Printing Office, February 1991).

Each study uses a wide variety of technical measures to achieve CO_2 emission reductions, but even similar conclusions for total CO_2 emissions in 2015—as shown by the National Energy Strategy, the Academy low-cost options, and OTA's moderate scenario—mask important differences in detail. For example, the National Energy Strategy appears to derive over half of its reductions in CO_2 emissions from supply-side changes—primarily increased use of nuclear power and biomass-derived transportation fuels. OTA and the Academy study get the bulk of their reductions from decreasing energy demand using technologies which, for the most part, exist today. We estimate that OTA and the Academy thus identify two to three times the low-cost energy conservation that the National Energy Strategy does. Both studies conclude that the buildings sector offers the largest potential for cost-effective emissions reductions, followed by industry.

In addition to these different assumptions about ease and cost-effectiveness of supply-side options versus demand-side options, there are differences between the policy measures presented in these analyses as options to achieve the technical reductions. The OTA and Academy reports discuss a mix of options including regulations (e.g., auto efficiency standards and building codes), financial measures such as taxes, federal RD&D, and information programs. In fact, the OTA study assumes that a mix of reinforcing options is crucial to the success of a reduction program. The National Energy Strategy relies extensively on the success of R&D programs and assisting the states through information programs. The only financial measures the DOE includes are incentives for energy production; regulatory programs that would force utilities to consider "end-use" efficiency are left to the states.

LOOKING TO THE FUTURE

The OTA report primarily addresses the next 25-year time frame, but Congress must not lose sight of the longer-term goals. Even under OTA's tough scenario, CO_2 emissions are projected to rise again shortly after 2015 unless new nonfossil energy supply can be brought on line before then. (This is because of the assumption that continued economic growth—demand for goods and services—eventually outstrips foreseeable prospects for further energy efficiency improvements.) To be able to make this transition to non-CO_2 sources (such as solar technologies or a new generation of nuclear power), a strong RD&D effort is needed—starting today.

Government can speed the process of testing and commercializing many energy-supply and end-use technologies. We note, however, that recent federal energy RD&D budgets are about half of what they were in 1980 and that only about five percent of the $3.7 billion national budget for energy technology RD&D in 1991 was devoted to renewable energy and conservation projects. We also note that any future interest in increasing these budgets will have to contend, at least, with the constraints imposed by the Budget Agreement and by the funding needs of other federal programs, both current and future.

We believe that OTA's moderate and tough cases are possible futures for the United States—but neither will happen without policy intervention. Again we stress that a combination of regulatory and market-based programs would be needed, a combination of performance standards, tax incentive programs, carbon-emission or energy taxes, labeling and efficiency ratings, and RD&D activities. The United States has little experience with such widespread change in energy usage patterns. We have limited evaluation of programs tried in the 1970s during the oil crisis; what those data suggest are that multiple instruments working toward the same goal are preferable. We note the breadth of these actions compared to the jurisdictions of any one committee of Congress—and the extent to which many of these governmental actions might affect our individual actions. And we see an unprecedented need for cooperation from the individual to the multinational scale if actions taken on behalf of global climate change are to be truly effective.

REFERENCES

1. U.S. Congress, Office of Technology Assessment. "Changing By Degrees: Steps to Reduce Greenhouse Gases," (OTA-O-482), U.S. Government Printing Office, Washington, DC (1991).

2. Solar Energy Research Institute, Idaho National Engineering Laboratory, Los Alamos National Laboratory, Oak Ridge National Laboratory, and Sandia National Laboratories. "The Potential of Renewable Energy, An Interlaboratory White Paper," prepared for U.S. Department of Energy, Office of Policy, Planning and Analysis, SERI/TP-260-3674, Solar Energy Research Institute, Golden, CO (1990).
3. Lawrence Berkeley Laboratory. "Buildings Energy Use Compilation and Analysis Project," Berkeley, CA (1986).
4. C.A. Goldman, Measured results of energy conservation retrofits in residential buildings," paper presented at *1986 ASHRAE Winter Meeting* (LBL-20950), San Francisco, CA, Jan. 19-22, 1986.
5. H.S. Geller. "Commercial Building Equipment Efficiency: A State-of-the-Art Review," contract prepared for U.S. Congress, Office of Technology Assessment, American Council for an Energy-Efficient Economy, Washington, DC (1988).
6. R.L. Ottinger et al. "Environmental Costs of Electricity," Oceana Publications (1990).
7. H.S. Geller. "Residential Equipment Efficiency: A State of the Art Review," contract prepared for U.S. Congress, Office of Technology Assessment, American Council for an Energy-Efficient Economy, Washington, DC (1988).
8. C. DiFiglio, K.G. DuLeep, and D.L Greene, Cost effectiveness of future fuel economy improvements, submitted to *Energy Journal* (1989).
9. C. DiFiglio, presentation at OTA workshop on transport and global warming, Apr. 6, 1989.
10. D.R. Bohi and M.B. Zimmerman, An update on econometric studies of energy demand behavior, *Annual Review of Energy* 9:105-154 (1984).
11. C.A. Dahl, Gasoline demand survey, *The Energy Journal* 7(1):67-82 (1986).
12. D.L. Greene, CAFE or price? An analysis of the effects of federal fuel economy regulations and gasoline prices on new car MPG, 1978-89, *Energy Journal*, 11(3):37-57 (1990).
13. B. McNutt and P. Patterson. "CAFE Standards—Is a Change of Form Needed?" SAE Technical Paper Series #861424, Society of Automotive Engineers, Warrendale, PA (1986).
14. S. Plotkin, U.S. Congress, Office of Technology Assessment, Increasing the efficiency of automobile and light trucks—a component of a strategy to combat global warming and growing U.S. oil dependence, testimony before the Senate Committee on Commerce, Science and Transportation, May 2, 1989.
15. A. Downs, The real problem with suburban anti-growth policies," *Brookings Review*, pp. 23-29 (Spring 1988).
16. D.L. Bleviss. "The New Oil Crisis and Fuel Economy Technologies," Quorum Books, New York, NY (1988).
17. U.S. Department of Energy, Energy Information Administration, "Indicators of Energy Efficiency: An International Comparison," EIA Service Report SR/EMEU/90-02, Washington, DC (July 1990).
18. S. Baldwin, Energy-efficient electric motor drive systems, *in:* "Efficient End Use and New Generation Technologies, and Their Planning Implications," T.B. Johansson et al., eds., Lund University Press, Lund, Sweden (1989).

19. M. Ross, Improving the energy efficiency of electricity use in manufacturing," *Science* 244:311-317 (Apr. 21, 1989).
20. F.W. Kirsch. "Energy Conserved and Costs Saved by Small and Medium-Sized Manufacturers: 1987-88 EADC Program Period," University City Science Center, Philadelphia, PA (March, 1988).
21. W. Fulkerson et al. "Energy Technology R&D: What Can Make A Difference? Volume 2, Supply Technology" (ORNL6541/V2/P2), Oak Ridge National Laboratory, Oak Ridge, TN (1989).
22. Electric Power Research Institute. "Power Plant Perform-ance Monitoring and Improvement," EPRI report CS/EL-4415, volume 3, Palo Alto, CA (February 1986).
23. National Association of State Foresters. "Global Warming and Forestry in the United States," background paper, Washington, DC (April 1990).
24. U.S. Department of Energy. "National Energy Strategy," First Edition 1991/1992, U.S. Government Printing Office, Washington, DC (1991).
25. Policy Implications of Greenhouse Warming—Synthesis Panel, Committee on Science, Engineering, and Public Policy, "Policy Implications of Greenhouse Warming," National Academy Press, Washington, DC (1991).

ENERGY EFFICIENCY, HUMAN ACTIVITIES AND CLIMATE CHANGE

Lee Schipper

Staff Scientist
Lawrence Berkeley Laboratory
Berkeley, CA 94720
and
Stockholm Environment Institute, Sweden[1]

Energy is important again, not because there is too little, but because of the threat that there is too much. Local and global environmental problems associated with energy use have increased their magnitude and threaten to grow even more as energy prices stabilize and demand grows even more. Of particular importance is the rise in CO_2 emissions from the use of fossil fuels that dominate energy use. How much will it cost to restrain their growth or even reduce them? Nothing at all, say optimists; too much, say pessimists.

This paper, the result of a two-year effort funded by the Stockholm Environment Institute with support from other programs at the Lawrence Berkeley Laboratory, examines how efficiency and other factors have changed energy use over the past twenty years, in order to think about the next twenty. We used a detailed, subsectoral analysis of the forces shaping energy use in the major industrial nations, the Soviet Union, China, and many developing countries. Using this analysis, we measured the achievements of energy efficiency to date. We then posed key questions that shape the answer to this question: how much might energy (or fossil fuel) demand grow; how much might energy-use efficiency improve; and how might these improvements be accelerated in ways that are economically beneficial?

Since 1973, there have been definite improvements in energy-use efficiency of between 2-3% per year, averaged over the U.S., Japan, Germany, and other European countries. Energy efficiency improvements can best be measured by examining changes in energy intensity, or the ratio of energy use to output produced (ie., energy used per tonne of steel, per kilometer driven, per square meter of home heated or cooled). Lower intensity generally means higher efficiency.

Using this definition, we can say the following:

Manufacturing has saved energy consistently, at the rate of over 2%/year before 1973 and at more than 3%/year between 1973 and 1985.

- The freight sector showed little progress, largely because of the increased energy requirements to move a tonne one kilometer by truck in most countries;
- Households show savings in both water or space heating, as well as electric appliances. Here both technology and behavioral changes (lower indoor temperatures) contributed. Heat needs in commercial buildings (offices, stores, hospitals) fell significantly. Electricity efficiency improved, too, but certain electricity uses, notably lighting, computing, and in many countries ventilation and air conditioning and heating increased significantly, offsetting much of the impact of significant efficiency improvements for electricity.
- Automobile efficiency changed slowly, especially outside the U.S. Indeed, the average car on the road in Europe used almost as much fuel per kilometer in 1990 as in 1973. In the U.S., by contrast, this figure fell 30%. Of course, the average size and power of cars driven in Europe increased significantly over that period.
- The efficiency of air travel increased markedly, as the requirements to fly a person one kilometer were halved, on average, in the OECD countries.

Overall, average energy intensity, as measured over the main end uses, in OECD countries fell about 20% between 1973 and 1985.

This is the good news. The bad news is that the rate of improvement of energy efficiency slowed in the mid 1980s, especially after oil prices crashed in 1986. As a result, present trends, while they portend efficiency improvements of close to 1% per year, are not strong enough to lead to the projected goals of 20% reduction in CO_2 production by 2010 as compared with 1990.

What are the prospects for improvements beyond the trends?

MANUFACTURING

The OECD increased its industrial output by 26% since 1970 (Figure 1) but reduced intensities and cut energy demand by 30%, mostly in reduced use of oil with little change in electricity or gas usage. We project that fuel intensity will continue to fall 2-3% per year with practically no change in electricity intensity.

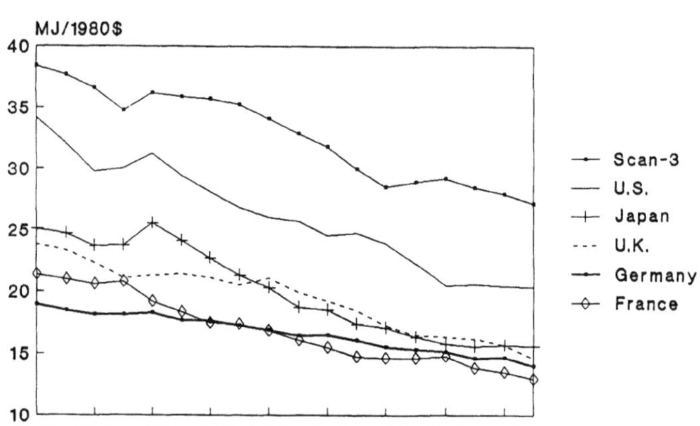

Figure 1. OECD Manufacturing Energy Intensities, Constant 1973 Industry Structure.

HOUSEHOLDS

Household use of energy is a key area of energy consumption. Existing patterns for space and water heating are grossly inefficient, but strides have been made and the potential is there to increase efficiency. Space heating efficiency (Figure 2) and appliance efficiency have improved markedly. New homes and appliances are 33% less energy-intensive than existing ones. Yet there remains a tremendous potential for improvement here. Even with our improvement, the U.S. is near the top in energy use for heating homes, largely because we heat more space per person than does Japan or countries in Europe.

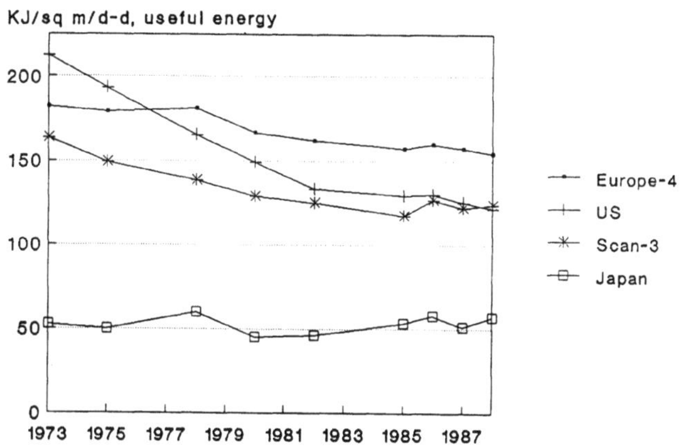

Figure 2. OECD Space Heating Intensities, 1973-1988.

TRANSPORTATION

There is trouble ahead in the area of transportation energy use. Aircraft travel is rapidly increasing. Although its energy intensity continues to decrease, energy for air travel is now rising. Truck freight efficiency has yet to show signs of reversing its poor record, which is really a result of increased use of trucks for just-in-time shipping, increased importance of smaller trucks, and increased congestion on roads.

Automobiles are a major problem. New car energy intensity in the U.S. has been rising slowly since 1986. The same figure of merit for new cars fell nominally in European countries, but showed little real gain when worsening traffic and driving habits are figured in. Cars bought in the U.S. still use about 33% more fuel/km than those bought in Europe (Figure 3). What about mass transit?

But overall auto efficiency did not improve very much anywhere except here in the U.S. (Figure 4). Figures show that we travel cheaper because of our gasoline prices. We travel the same distances as do Europeans but we travel more often (Figure 5). Bear in mind also that Europe has one-third fewer cars per capita than we have. Europeans drive faster; there are essentially no enforced speed limits, but fuel costs are high. Sweden raised its gasoline price by $1.40 per gallon in 1990. We refuse to raise our gasoline taxes and Europe refuses to lower its speed limits!

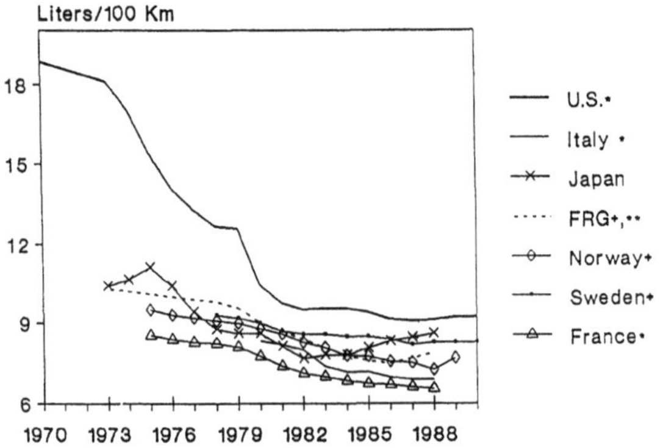

Figure 3. New Automobile Fuel Economy.
*Includes diesels; U.S. includes light trucks.
+Excludes diesels.
**"Test values."

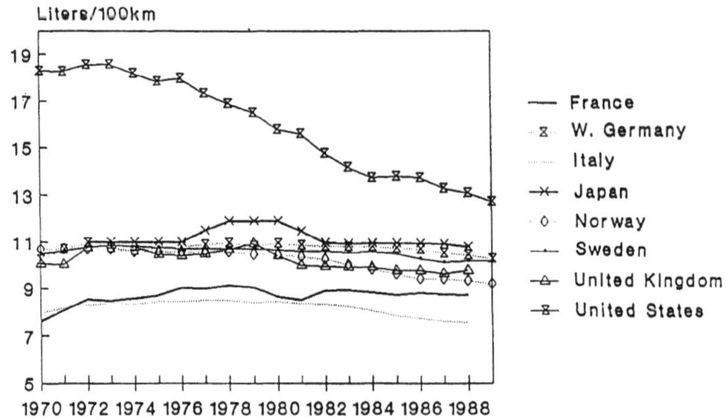

Figure 4. Automobile Fuel Economy, on Road (Actual) Fleet Averages. Includes diesels (except Sweden and Norway), and personal light trucks. (U.S., Germany 1989 preliminary.)

Today, OECD countries are not gravitating toward smaller cars and more efficiency. The problems are not mainly technological; they are nontechnical. Future energy use by cars depends on gasoline prices and on the attitude of the population toward speed limits and the size of cars. We must accept the 50 mile-per-gallon car.

Busses and trains carry only about 4% of U.S. travel, versus typically 15% in Europe. Differences in fuel prices are one important reason. Our gasoline is so cheap that we do not need to care about fuel costs as much as people in Europe. At the same time, the automobile's share of travel kilometers continues to increase in all countries both because the number of cars is catching up to the U.S. level and because average yearly driving distances are slowly creeping up.

Comparing gross domestic product (GDP) to energy use is a very coarse comparison. In all countries "lifestyle" factors such as homes, cars and appliances make a big difference as does the manufacturing/production mix. Energy intensity differences are real. The U.S. has the worst efficiency in manufacturing and automobile use. Only new 1993 appliance standards will pull us up even in that area. In sum, we are the most inefficient and have the highest energy-to-GDP ratio. Seen from abroad, we are heating 50-75% more area than the Europeans or Japanese and we travel about 50-75% more than the Europeans. We represent to the world a unique combination of cheap fuel and urban sprawl, but can we continue these practices under a no-regrets policy? Such differences also should be kept in mind when making international comparisons of "efficiency". That is, more than half of the difference between American and Japanese energy use per unit of gross domestic product can be traced to Americans' greater indoor space and travel. Outside of differences in industrial energy-use efficiency, where Japan has a clear advantage, the Japanese edge in efficiency is actually relatively small.

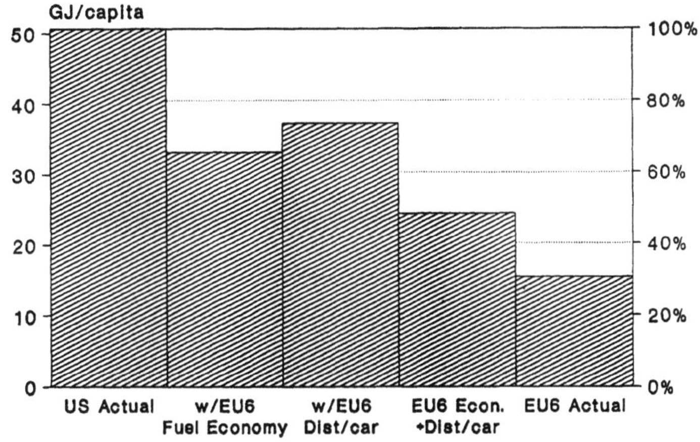

Figure 5. Per Capita Energy Use by Automobiles, Comparison of U.S. and Europe* in 1987. *W. Germany, France, Italy, U.K., Norway, Sweden, weighted by actual consumption and distances driven.

Looking at the future of wealthy countries under a variety of scenarios for the year 2010 (Figure 6), we can see already available efficiency improvements applied to the present system resulting in lower CO_2 emissions per unit of activity, or per unit of gross domestic product. But increases in either measure raise overall CO_2 output from the wealthy countries. (Of course, massive substitution of natural gas or nuclear power for coal, or renewables for any fossil fuels, could lower CO_2 output drastically, but such changes are unlikely under most assumptions of energy prices and policies during the period to 2010.)

Our "Moderate" scenario, under which a greater efficiency effort is undertaken (and in which energy prices lie roughly 25% higher than they do today, largely to account for the clean up of land and air pollution), could just about hold energy use in 2010 to its 1985 level; CO_2 emissions could decrease from this figure with a modest effort of substitution away from fossil fuels, principally from coal. The efficiency efforts in this scenario are constructed by assuming that the average energy efficiency in 2010 is slightly better than the best new efficiency in 2010. (Important exceptions are trucks and autos which improve considerably more.)[2]

We also designed a "vigorous" policy scenario that is supported by a broad consensus that the greenhouse effect portends many significant direct and hidden costs if not mitigated by strong actions. These include international agreements on energy efficiency improvements of major energy-using consumer goods, accelerated research and development of both improved efficiency and non-CO_2 producing energy sources, roughly doubled energy prices compared with 1990 levels, and partly as a consequence of these stimuli, some "lifestyle" changes. The most important of these include reduced travel, more of which is carried out by rail or bus. Under these assumptions, CO_2 emissions would drop 55% relative to their 1985 level, even before any changes in input fuels were considered.

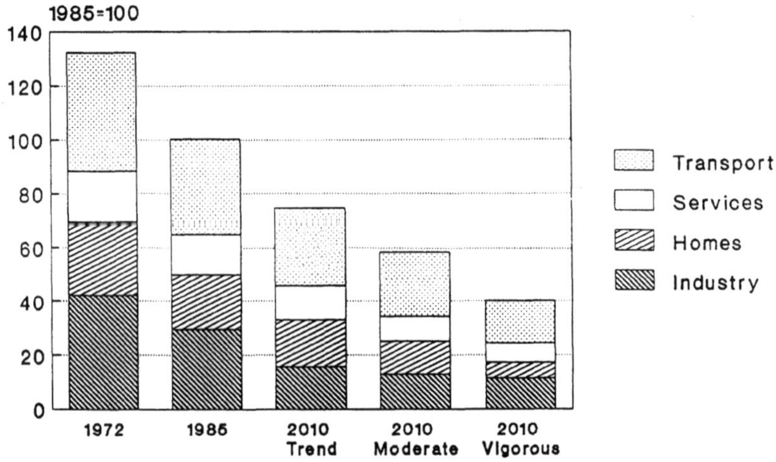

Figure 6. OECD Final Energy Demand in 2010, Effect of Intensity Reductions.
Note: Structure and activity held constant at 1985 levels.

In the real industrialized world, however, overall population and GDP per capita are increasing, and travel, home comfort items and services are growing faster than GDP. In our scenarios, we estimated hat these changes alone would raise energy use by nearly 80% over its 1985 level. This is why the "Trends" scenario still foresees greatly increased CO_2 emissions, while emissions in the "Moderate" scenario are barely held to their 1985 levels. Combining this growth with the bold improvements foreseen in the "Vigorous" scenario gives results that lead to clear reductions in CO_2 over 1985 levels.

It is because of this natural growth in the system that I became convinced that relying on technological fixes alone to improve energy efficiency will not lead to reductions in CO_2 emissions that could meet the goals of a 20% reduction compared with 1990, if such a goal were itself deemed worth while. The problem is not that such technological fixes would not in principal reduce fossil fuel use over what it otherwise would be. Rather the problems are that 1) we cannot say with certainty which technologies will be implemented (nor how rapidly) and 2) we cannot reasonably believe that the efficiency improvements over the next two decades would be large enough to also compensate for the forces increasing CO_2 emissions. In other words, some lifestyle changes must be on the table. This is particularly true for Americans, who, seen in an international light, enjoy far more energy intensive services (space heated or cooled, distances driven) per unit of GDP than Europeans.

But many of our present practices and policies, such as company cars (a European phenomenon) and home interest income-tax deductions (a perk in the U.S. and some European countries), which stimulate energy use, should be challenged. That is, if one is concerned that climate change stimulated by the greenhouse effect poses a significant threat to future generations, then some of the pleasures of present generations should be at least questioned, if not curtailed through market forces. Indeed, one of the advantages of a CO_2 tax, as opposed to other measures to stimulate reduced CO_2 emissions, is that consumers are free to balance their responses between lifestyle changes (i.e., less driving, smaller homes), behavior changes (smaller cars, lower indoor temperatures), and efficiency improvements (more efficiency cars, better thermal insulation).

Putting all these options on our own agenda will also have an important effect on how others in the world view the industrialized countries' response to climate change threats. Take the formerly planned economies of Eastern Europe. The reforms in these countries will probably lead to enormous efficiency increases, which, together with reductions in coal use to improve air and water quality, should lead to significant decreases in CO_2 production through the early part of the next century. But the improvements in living standards in the developing countries will undoubtedly lead to increased CO_2 emissions, even with greatly improved energy efficiency. Only if the OECD exerts leadership both in technological improvements and in behavior and lifestyle will Third world countries take our actions to mitigate climate change seriously, work to improve their own energy efficiency (perhaps with our help), and perhaps avoid some of the lifestyle patterns we in the industrialized countries developed that have increased fossil fuel use so much.

Admittedly many questions remain unanswered by this brief analysis. Does equity demand that the wealthy world do more than the others? If so, what "lifestyle" changes will be necessary? Let us suggest some strategies to improve energy efficiency. We need proper private pricing of energy to include marginal costs. Costs of energy must include environmental costs. People must become aware of and sympathetic to energy use and efficiency, and we need strong government leadership. Building standards must be raised. Motor vehicle efficiency and emissions must improve. We must accelerate technology transfer to non-OECD countries.

Above all, we must have honest expectations about efficiency programs and what they can deliver. World populations must be behind the movement to improve. With this in mind, we need to eliminate all indirect subsidies to energy use. We must restrain automobile use and reduce the need to travel by lessening urban sprawl and using new technology in communication. Above all, we need to restrain worldwide population growth.

Present U.S. policy leaves little consolation for our children and grandchildren. Few of our leaders accept the seriousness of environmental and climate challenges. In that strange sense, the Administration's posturing up to the U.N. Conference on Environment and Development in Rio in June 1992 may have been a realistic reflection of the lack of real motivation among Americans for far-reaching environmental change. After all, the European countries to which America is often compared have moved quickly to establish at least moderate taxes on CO_2 and discussed other measures of a stronger nature that might limit future CO_2 emissions. But American politicians are afraid to raise energy prices. American industry has formed various lobby groups to work against CO_2 taxes, a strange alliance of so-called free-marketeers against using market forces! Sadly, some environmentalists pay lip service to the pricing issue, claiming that it is a "free lunch." The "free lunch" efficiency policies may have been oversold, and represents a convenient way to pass the problem to future generations. Future generations have no standing in court. Would you trust their environment to lawyers and lobbyists?

In conclusion, the costs of restraint of CO_2 emissions depend upon the changes required. Moderate reductions in CO_2 are possible and the cost may be a negative one but much work needs to be done. Market forces alone will not solve the problems, but without them, nothing else will. Long-run predicaments and solutions are human.

NOTES

[1]Opinions strictly those of the author and not of the Lawrence Berkeley Laboratory, the Stockholm Environment Institute, or any of their sponsors.

[2]Details of these scenarios are given in our book, *Energy Efficiency and Human Activity: Past Trends, Future Prospects* by Lee Schipper, Steve Meyers, et al., (Cambridge University Press, 1992).

THE COSTS OF REDUCING U.S. CO_2 EMISSIONS: FURTHER SENSITIVITY ANALYSES[1]

Alan S. Manne

Professor
Stanford University
Stanford, CA 94305

Richard G. Richels

Director
Energy Analysis and Planning Department
Electric Power Research Institute
3412 Hillview Avenue
Box 10412
Palo Alto, CA 94303

INTRODUCTION

In a previous paper, we used the Global 2100 model to explore the implications of a carbon constraint upon domestic energy costs and the resulting effects on the U.S. economy as a whole (Manne and Richels, 1990). The impact of a CO_2 limit will depend on the technologies and resources available for meeting demands as well as on the demands themselves. Given the enormous uncertainty surrounding these factors, losses were calculated under alternative assumptions about each.

The news for policy makers was mixed. The bad news was that if emission controls are required, there will be substantial costs, perhaps as large as several trillions of dollars over the next century. The good news was that the size of the ultimate bill can be significantly reduced through successful R & D in both the supply and the demand sides of the energy sector.

There have been a number of critiques of our analysis. Although some reviewers feel that we have been unduly pessimistic in estimating the costs of a carbon constraint (e.g., Williams, 1990), others have suggested that we may have erred on the low side (e.g., Hogan and Jorgenson, 1990). The differences arise primarily from alternative views about the potential for supply and demand side improvements in the energy sector. At issue for policy making is the impact of such

Global Climate Change, Edited by J.C. White
Plenum Press, New York, 1992

differences on the overall cost of a carbon constraint. In this note, we expand our earlier analysis to encompass a broader range of views about the future character of the energy sector. We also explore a more ambitious carbon limit than that considered in the original analysis. As in the earlier paper, we focus on the impacts of a carbon limit upon the cost of energy to the U.S., and do not attempt to analyze the benefits of a global carbon agreement.

KEY DEMAND PARAMETERS

Energy consumption need not grow at the same rate as GDP. Over the long run, they may be decoupled. In Global 2100, these possibilities are summarized through two macroeconomic parameters. One is abbreviated ESUB (the elasticity of price-induced substitution), and the other is AEEI (autonomous energy efficiency improvements). If there is sufficient time for the adaptation of capital stocks, most analysts will agree that there is a good deal of possible substitutability between the inputs of capital, labor and energy. The degree of substitutability will affect the economic losses from energy scarcities and price increases. In our aggregate model, the ease or difficulty of these tradeoffs is summarized by ESUB. The higher the value of ESUB, the less expensive it is to decouple energy consumption from GDP growth during a period of rising energy prices. When energy costs are a small fraction of total output, ESUB is approximately equal to the absolute value of the price elasticity of demand. In Global 2100, this parameter is measured at the point of secondary energy production—electricity at the busbar, crude oil and synthetic fuels at the refinery gate—and is taken to be 0.40. That is, a 1% price increase will lead to a decline of 0.4% in the demand for energy.

In addition to the reductions in energy demand brought about by rising energy prices, there is also the impact of autonomous energy efficiency improvements (AEEI). Such non-price induced efficiency improvements may be brought about by deliberate changes in public policy, e.g., mandatory fuel efficiency standards for automobiles. They may also occur as a result of shifts in the basic economic mix away from manufactured goods and toward more services.

Although this is a "top-down" method of analysis, it does not automatically lead to quantitatively different conclusions than those obtained through a "bottom-up" approach. Both methods must deal with a historical record that contains limited amounts of evidence on the role of price-induced versus autonomous conservation. Clearly, Williams interprets the statistical record in a very different way than Hogan and Jorgenson. Our values of ESUB and AEEI have been chosen so that the Global 2100 energy projections will match with the conventional wisdom expressed in the median responses to the International Energy Workshop polls reported by Manne and Schrattenholzer (1989).

Several critiques of our earlier analysis emphasized the importance of the AEEI assumption. Figure 1 shows the effect of alternative parameter values. With autonomous energy efficiency improvements occurring at an annual rate of 1.5%, by the end of the next century the economy would require only 20% the amount of energy that it would with an AEEI of zero. This is entirely apart from the conservation that would be induced by rising energy prices in response to the depletion of conventional oil and gas resources.

Clearly, the AEEI parameter is highly controversial. To represent distinct viewpoints, we will explore values ranging from zero to 1.5% per year.

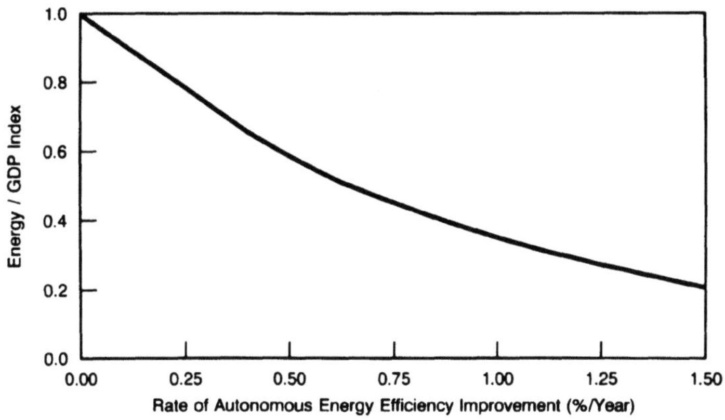

Figure 1. Effect of AEEI on Total Primary Energy Consumption - 2100.

SUPPLY AND COST ASSUMPTIONS

A second determinant of the costs of a CO_2 limit is the availability of low-cost non-carbon based supply alternatives. If such options were abundantly available, there would be little or no cost to reducing emissions. Indeed, reductions would be induced automatically through the workings of the price mechanism. Unfortunately, there is considerable disagreement regarding the cost and availability of carbon-free alternatives to fossil fuels.

In our earlier study, we confined our supply side sensitivity analysis to the electric sector. Scenarios were constructed around the availability of a low-cost carbon-free electric generating technology. Although the cost and performance characteristics were similar to those for an advanced nuclear design with passive safety features, any number of carbon-free technologies may ultimately turn out to be economically competitive. For this reason, we referred to the advanced low-cost technology simply as ADV-LC.

Table 1 identifies the alternative sources of nonelectric energy that are included in Global 2100. With regard to carbon-free alternatives, the choices have been grouped into two broad categories: RNEW (low-cost renewables such as ethanol from biomass) and NE-BAK (high-cost backstops such as hydrogen produced through photovoltaics and electrolysis). The key distinction is that RNEW is in limited supply, but that NE-BAK is available in unlimited quantities at a constant but considerably higher marginal cost.

A criticism of our earlier work was that we had not allowed for substantial quantities of energy sources falling into the RNEW category. In the following sensitivity analysis, we investigate how the costs of a carbon limit would change with the availability of low-cost renewables. Specifically, we explore scenarios with annual RNEW limits of 0, 10, and 20 quads. (These limits might be associated, for example, with alternative assumptions on the quantities of land and water that are available for the cultivation of biomass.) As a point of reference, oil imports were running at the annual rate of 15 quads in 1990, and total primary energy consumption was 82 quads.

Table 1. Nonelectric Energy Supplies

Technology name	Description	Carbon emission coefficient, tones of carbon per million BTU of crude oil equivalent[1]	Unit cost 1988 $ per million BTU of crude oil equivalent
OIL-MX	Oil imports-exports	.0203	3.30 in 1990, rising to 15 from 2030 onward
CLDU	Coal—direct use	.0251	2.00[2]
OIL-LC	Oil—low cost	.0203	2.50
GAS-LC	Natural gas—low cost	.0145	1.50[2]
OIL-HC	Oil—high cost	.0203	6.00
GAS-HC	Natural gas—high cost	.0145	5.00[2]
SYNF	Synthetic fuels	.0408	10.00
RNEW	Renewables	.0000	9.90
NE-BAK	Nonelectric backstop	.0000	20.00

Notes: 1 Source of carbon emission coefficients: Edmonds and Reilly (1985).
2 To allow for burner-tip equivalence, an additional $1.25 per million BTU is added to allow for gas distribution costs.

SENSITIVITY ANALYSIS

We begin with the same carbon limit that was explored in the earlier paper. That is, the absolute level of emissions is held to its 1990 level through 2000, gradually reduced to 80% of this level by 2020, and then stabilized thereafter. Unlike Williams, we do not rely upon "technology forcing." That is, we assume that the economy is always operated so as to achieve its best performance—both with and without a carbon constraint. Aside from its impact upon global climate change, this means that a carbon constraint cannot improve economic performance in any one country. Inevitably, there are losses—here expressed in terms of the discounted value of conventionally measured macroeconomic consumption.

Figure 2 extends our earlier work in two ways. The losses are presented for a broader range of assumptions about the potential for supply enhancements and demand conservation. To keep the number of scenarios manageable, we assume that the low-cost carbon-free electric technology, ADV-LC, is available in all cases. For purposes of comparison, the left most bar in Figure 2 is identical to scenario III in Manne and Richels (1990).

It turns out that the direct economic losses are quite sensitive to assumptions about both demand and supply factors. On the demand side, the principal issue is the rate of autonomous energy efficiency improvements. On the supply side, the principal controversy concerns the availability of low-cost, carbon-free alternatives to oil and natural gas. For the losses to approach zero, however, the most optimistic combination of supply and demand assumptions must be adopted.

An AEEI of 1.5 is required to reduce nonelectric energy demands to the level at which they can eventually be covered by 20 quads per year of RNEW, the low-cost renewable supply technology. Since we have assumed that RNEW is competitive with the coal-based alternatives and since no costs are imputed to autonomous energy efficiency improvements, there are negligible total losses accruing under this scenario.

We doubt that an AEEI of 1.5 is sustainable over a long period. This would exceed the U.S. GDP growth for much of the 21st century. In turn, this would mean that a rising GDP is accompanied by falling energy demands during many decades of stable energy prices. Some critics of our earlier analysis argue that because we only considered AEEI values as high as 1.0, we were overly pessimistic about the potential for energy efficiency improvements. It is important to note that a low autonomous rate does not necessarily imply a low overall rate of energy efficiency improvements. In addition to the AEEI effect, there is also the effect of rising energy prices on demands. One must analyze both of these effects together in order to determine the degree of decoupling between energy consumption and GDP growth.

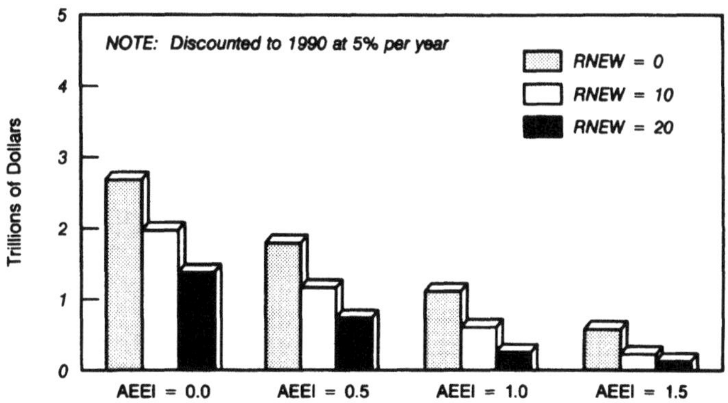

Figure 2. Discounted Costs to USA, 20% Reduction in Carbon Emissions.

Consider, for example, an AEEI of 0.5. Figure 3 shows the projected growth rates for energy consumption and GDP during the period between 2000 and 2020, a time when the economy will have the most difficulty in adjusting to a 20% reduction in carbon emissions. Without a carbon limit, the GDP and energy consumption grow at annual rates of 2.3% and 1.0%, respectively. The AEEI accounts for less than half the difference. The remainder is due primarily to consumer response to the higher prices of oil and natural gas as these resources undergo gradual depletion both in the U.S. and abroad.

With a carbon limit, there is a high implicit tax on carbon, and the price-induced effects become even more pronounced. Energy and GDP are virtually decoupled during this period—with the price mechanism accounting for more than three-quarters of the reduction in energy use. Figure 3 tells us that even with an AEEI of 0.5%, the overall rate of energy efficiency improvement may be 1.5% or more. Unlike autonomous improvements, the contribution from price-induced conservation is not costless. There is no free lunch.

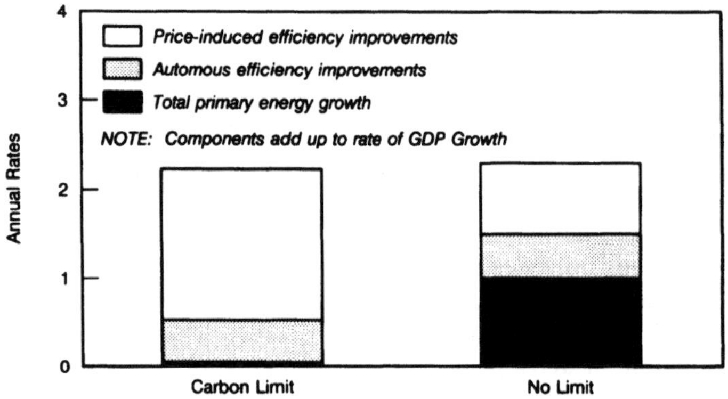

Figure 3. Decoupling Between Energy Consumption and GDP Growth, 2000-2020.

CARBON TAXES

In Global 2100, we make no attempt to quantify the losses due to inefficient regulatory behavior in the administration of a carbon limit. Instead, it is supposed that uniform carbon taxes would be levied so that emissions are allocated optimally across all sectors of the economy. The carbon tax rate may then be interpreted as the marginal productivity of carbon at different points of time. This is to be distinguished from the average productivity of carbon—as measured in terms of consumption of GDP losses.

If a carbon limit is deemed desirable, there are persuasive economic efficiency arguments for employing a carbon tax in order to decentralize implementation. There is, however, considerable controversy over the level of the carbon tax that would be required to induce the desired shift away from carbon intensive fuels. In our earlier paper, it turned out that the long-run equilibrium tax was fixed at a level that would make synthetic fuels (SYNF) and the nonelectric backstop (NE-BAK) technologies equally attractive. Specifically, the equilibrium tax was determined by the ratio of their cost differential to their carbon coefficient differential:

$$\frac{\text{Cost differential,\$/mmBTU}}{\text{Carbon differential,tons/mmBTU}} = \frac{\$20\text{-}\$10}{0.04 \text{ tons}} = \$250/\text{ton}$$

It has been suggested that our long-run equilibrium tax level was too high because we were unduly pessimistic about the availability of low-cost, carbon-free alternatives to oil and natural gas. If such supplies were available in sufficient quantities, they would replace the nonelectric backstop as the marginal source of supply, and would therefore lower the level of the equilibrium tax. The introduction of the RNEW technology provides us with a convenient means of exploring this issue.

Figure 4 shows the time paths for the carbon tax under our most optimistic assumptions about RNEW. This is the case in which we assume that up to 20 quads/year may be supplied by this technology at costs which are competitive with coal-based synthetic fuels. The overall effect is to delay the time at which the backstop technology (NE-BAK) becomes the marginal source of supply. The economy will always use low-cost renewables first. It is only after this source is pushed to its limit that it will pay to turn to the high-cost nonelectric backstop, NE-BAK. With an AEEI of zero, this happens in 2040. With higher AEEI values, there are lower overall demands for energy, and the introduction of the backstop is delayed further into the future. With an AEEI of 1.5, the high-cost backstop is never needed, and it becomes irrelevant to the determination of the equilibrium carbon tax.

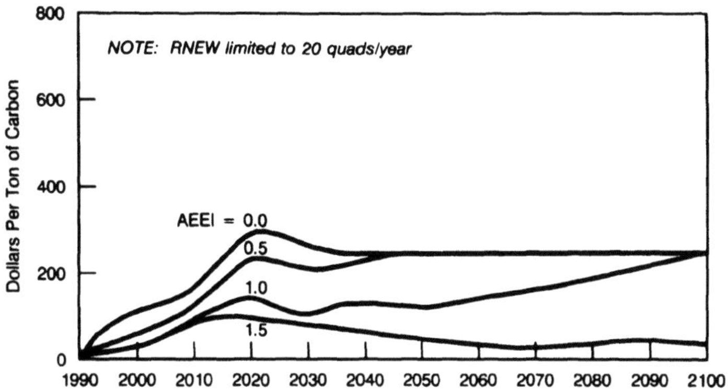

Figure 4. Carbon Taxes Under Alternative AEEI Assumptions, 20% Emissions Reduction.

A TIGHTER CARBON CONSTRAINT

Our final calculations refer to an alternative carbon constraint. Instead of reducing carbon emissions to 80% of their 1990 level, we assume a more severe cutback—a 50% reduction. Proposals of this drastic nature are under discussion. It is important to understand their implications—both in terms of benefits and costs.

In the following calculations, we assume that emissions are reduced by 20% below 1990 levels by 2000 and 50% below these levels by 2020. Figure 5 shows the losses that would accrue from this constraint—again under alternative assumptions with respect to the AEEI and the quantity available of low-cost renewable nonelectric energy. The tightening of the carbon limit leads to a significant increase in the absolute level of losses, but the general appearance of the sensitivity analysis looks similar to Figure 2. In order to reduce losses to negligible levels, we must again make optimistic assumptions with respect to both the AEEI and RNEW parameters. There is no easy path to a low-carbon future.

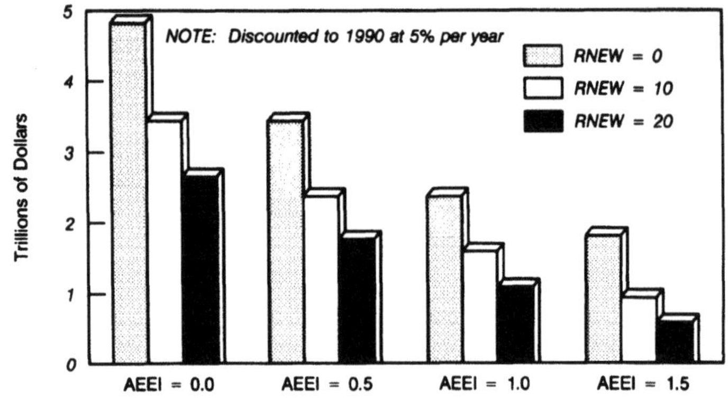

Figure 5. Discounted Costs to USA, 50% Reduction in Carbon Emissions.

ACKNOWLEDGEMENTS

The research reported in this paper was funded by the Electric Power Research Institute (EPRI). The views presented here are solely those of the individual authors, and do not necessarily represent the views of EPRI or its members. The authors are much indebted to Diane Erdmann for research assistance. We have benefitted from discussions with: Robert Dorfman, George Hidy, William Hogan, Dale Jorgenson, Henry Lee, David Montgomery, Stephen Peck, Lee Schipper, John Weyant, Robert Williams and David Wood.

NOTES

1. Reprinted with permission from *The Energy Journal*, Volume 11, Number 4, 1990. All rights reserved.

REFERENCES

Edmonds, J., and Reilly, J.M., 1985, "Global Energy—Assessing the Future," Oxford University Press, New York.

Hogan, W.W., and Jorgenson, D.W., 1990, "Productivity Trends and the Cost of Reducing CO_2 Emissions," Harvard University, Energy and Environmental Policy Center, Cambridge, MA.

Manne, A.S., and Richels, R.G., 1990, CO_2 emission limits: An economic cost analysis for the USA, *The Energy Journal* (April) 11:2.

Manne, A.S. and Schrattenholzer, L., 1989, The International Energy Workshop—A progress report, *OPEC Review* (Winter).

Williams, R.H., 1990, Low-cost strategies for coping with CO_2 emission limits, *The Energy Journal* 11:4.

DISCUSSION

A SUSTAINABLE ENERGY FUTURE—TECHNOLOGICAL OPTIONS FOR CHANGE

QUESTION: Have you factored in a possible greatly increased number of low, very low and zero emission vehicles that might be required to meet the Clean Air Act mandate?

FRIEDMAN: First of all, I should clarify that the low, very low and no emissions refers to hydrocarbons for controlling ozone. This is not carbon.

The results that I presented, in fact, assumed no penetration of alternatively fueled vehicles, but we do explore such a scenario in our assessment. Let me just point out a couple of reference points.

If the alternatively fueled vehicle—the low-emission vehicle—takes hold and is 85% methanol, we would calculate that there is virtually no change in carbon emissions from a gasoline vehicle to an M-85 vehicle. If we go to improved methanol vehicles, vehicles that can use 100% methanol, there might be about a 15% improvement in greenhouse gas emissions from alternatively fueled vehicles.

For natural gas we would also assume about a 15% improvement. If the alternative vehicles are electric vehicles, it would obviously depend on what was used to generate that electricity. If that electricity is from a new coal plant, there would be a 5% increase in the CO_2 emissions as compared to gasoline. If it is from a state-of-the-art gas turbine, it might be about a 45% decrease. That will give you some idea of the impacts of alternatively fueled vehicles on CO_2 emissions.

QUESTION: Comparing Europe to the U.S., could Europeans further conserve energy by enforcing speed limits? How does the speed limit play into the European scenario?

SCHIPPER: We all have sacred cows—speed limits in central Europe, company cars in Britain, and low gasoline prices here.

I was at an interesting meeting in East Berlin on transportation, sponsored by the IAO and OECD. A Frenchman showed a beautiful slide that showed that the test fuel consumption at 50 miles an hour of a car designed to go 150 is terrible, and the Germans jumped up and took their shoes off and stamped on the table. They asked, "How dare you even imply that we might want speed limits!" The European no-limit speed laws are just as culturally insane as our fixation with low gasoline prices.

Will the Germans lower their speed limits? Well, when the Italians did, the Germans protested. They couldn't drive fast in Italy. The Italians have a speed limit, but when Fiat sent a car to pick me up to discuss speed with them, we were going 170 kilometers per hour. The speed limit was 110 or 120 and it had just snowed.

There is a cultural problem with speed in Europe that is one of the reasons why the European mile-per-gallon curves haven't improved. More and more are going faster and faster. They have a different valuation of human life behind the wheel. When two cars going at German, Austrian, Italian or French speeds crash, they basically fuse. And it is really very scary—one, because it happens and, two, because the speed limit issue in Germany is like the abortion or the gun issue in this country.

QUESTION: Literature has often cited a 1-5% GDP loss from a 20% CO_2 reduction from your model. Today you put the range from 0-3% GDP loss. Have you reduced your cost estimates?

RICHELS: In the original analysis that we conducted for the United States, the 5% number was based on what we call a highly constrained supply-and-demand scenario, and that scenario is often misinterpreted as to what we were doing. We constructed the highly constrained supply-and-demand scenario in order to calculate the benefits of successful R&D on both the supply side and the demand side of the energy sector. We constructed a scenario from which we could measure the benefits of nuclear power, the benefits of more efficient coal technologies, of CO_2 scrubbing, of greater efficiencies on the demand side. In the paper, we stated that we did not imply that this was what the future was going to look like. This was a necessary baseline—not a base case but a baseline—for the calculation of benefits from R&D.

That 5% number has been somewhat misinterpreted. I would call it the maximum cost of a 20% reduction in carbon limits by the year 2020. If you want to get a 20% reduction earlier than 2020 or if you want to get greater than a 20% reduction in carbon emissions, such as a 50% reduction in carbon emissions, 5% could be quite operative.

In fact, we have been criticized by some analysts—particularly Bill Hogan and Dale Jorgenson—in being somewhat optimistic as far as the upper bounds on the cost of the carbon constraint. As Howard Gruenspecht pointed out to me yesterday, we assume that the economy works optimally, efficiently, all markets are cleared and so forth, to the extent that we behave irrationally and suboptimally. Again, our estimates could be low. The numbers could be higher or lower and there are strong cases that could be made in either direction.

GLAUTHIER: Mr. Schipper, do you want to comment on that?

SCHIPPER: One point in Rich's behalf. Rich and I attended a workshop at MIT in the summer of 1989 organized by the late David Wood, who was trying to get at this question, and Peter Passle from *The New York Times* got some of the papers. Suddenly in November there were these articles and the cat was out of the bag and the world was going to go to hell. Now they didn't ask Mr. Richels, "What do these numbers mean?"

I don't mean to say that preliminary stuff shouldn't be circulated, but this terrible boiling controversy grew up while they were constantly doing runs and reruns. Those curves indicate a 0-3% reduction in GNP. There is no intent to deceive anybody. It may not cost anything at all. It may cost a little.

The crucial question that neither Rich nor I can answer is: is 3% a lot, is 1.5% a lot? That is the values question. And *The New York Times* didn't go at that either, and as a result Rich and I spent two years, in a sense, fighting a battle that never should have been declared in session, because these numbers just don't get understood. They don't have the meaning that we would like to assign to them when we put them into editorials.

QUESTION: What is the role of the states versus the federal government in encouraging efficiency, the use of alternatives, carbon taxes, et cetera?

FRIEDMAN: I think the states can play a very large role and, in fact, do play a very large role right now and within two sectors.

First, in the way they deal with the electric utility sector. State public utility commissions regulate the ways that utilities make profits. It is up to the state public utility commission to decide whether, in fact, they will allow their utilities to profit from efficiency investments, certainly one very important area.

Some states, or at least California, are beginning to experiment with transportation incentives. There have been discussions of a rebate program, which could certainly be a federal or a state option. Efficient cars in California might receive a rebate from money raised from taxes paid on less efficient vehicles.

Certainly in the area of transportation controls and planning, the state and the local governments are very active. So I believe—but then, again, I work for the Congress—that the federal government must take a very active lead in this if we are to have consistent policies that help out industry.

For all these reasons I believe that there should be federal involvement, but there are lots of actions that can be taken at the state and local levels.

SCHIPPER: I think that California, for better or for worse, is doing a lot of the right things. The "for better" part is that we are doing something and we are the sixth largest economy in the world.

Last year there was a meeting called "Ecology '90" in Göteborg, and the Swedish attache for technology in Los Angeles (LA) arranged a session on the LA air problem. That was the best attended session in the entire conference. Why? Because Volvo, Saab and all the auto companies realize that California standards will probably be their future.

I think that the Los Angeles plan is crazy in the sense that it says you have to drive an electric car, you can't use your barbecue, you have to stay home. It is really easy to say there is a huge externality in driving in LA. We are all going to pay more. But, of course, no one wants to confront that.

At least, though, California is trying something and it is going to have a national impact. The problem is that the crybabies are saying that it costs business more to deal with these problems. That is part of the market-based solution, whether it is forced through regulation or directly through taxes, but you are going to hear everybody trying to play off the fact that there are no carbon taxes in Arizona or Nevada. The down side of this is that one state does something and then the crybabies—and they are on all sides—threaten to get up and move away. So that does throw some of the responsibility back from the state to the federal government.

Let me add one other point. The Scandinavians, the Dutch, the Swiss are acting alone, because I think they realize, if we charge ourselves, our economies and our industries will respond. The Americans and the Japanese are saying, no, we can't do this unilaterally. So you have the issue that will come up in the discussion tomorrow which is, what if the developed nations do something? Do they have to wait for the developing countries to sign on or not?

My view is, we shouldn't wait for them. We have to start. But again, there are all kinds of cries that will come up. This will be a regional issue. No matter where you start to act, someone will say, "But at a higher level they are not acting." On the other hand, you can't get agreement at the higher level.

In the end, if there is an ethical problem here, it has to start where those people agree openly to begin, as in Scandinavia or California, and then hope the others will catch on. In the case of clean air, the northeast is signing on with California.

QUESTION: Everyone talks about a drop in GDP when carbon constraints take effect. Do the estimates include the jobs created in new manufacturing and operating procedures vis-a-vis the shift to a service economy?

RICHELS: No, and I believe that the next stage of analysis has to be more sectoral, looking at the cost impacts sector by sector and looking at both the positive and the negative implications.

QUESTION: Presumably you used a cost criteria for inclusion of certain measures in your moderate and tough scenarios. What was the marginal cost of reductions in the moderate and tough scenarios?

FRIEDMAN: The way we decided what particular technical option fell into which scenario, "moderate" or "tough," was in workshop formats.

These were general consensus decisions rather than establishing strictly cost criteria. Our "moderate" scenarios we call "cheap and easy," not cheap or easy. So it had to be not only a life-cycle cost-effective judgment, on the order of a 6% discount rate (and these are simple compliance costs minus fuel savings), but also a subjective judgment of how much individual pain might be involved with actually adopting such a technology or, in some cases, lifestyle change. So all of those that we rather subjectively, put in this group of "cheap and easy" scenarios ended up in the "moderate" category. All others ended up in the "tough" category.

Where did we cut off "tough" scenarios? It wasn't based on a marginal cost. I have been using Richels' terms—it wasn't a theoretical potential for energy efficiency. It was more a technical potential. What is the best that is out there now, what can we point to that is a type of a technology that we can be assured is going to be around? So our "tough" category did not have a marginal cost associated with it and some of the marginal costs are quite high.

SCHIPPER: One of the things that is unique to America is that lawyers deal with these enormous cost calculations in hearings. We are fooling ourselves if we think we have four-digit accuracy as to what the cost of saved energy will be in Vermont for a two-door refrigerator that has through-the-door frost-free ice, yet that is the way policy is made.

Our guess is that in our moderate scenario it is $500 to $1,000 per house just for the retrofitting and another $1,000 into the vigorous scenario. And the economics of the moderate scenario aren't bad. It is about $15 per barrel of oil saved. For "vigorous" it is a lot more because you don't save as much and you pay just as much.

For automobiles, I am willing to accept a compromised cost of saved energy. I don't believe the methodology makes sense anymore because anything that saves fuel in automobiles also boosts power. Nevertheless, I think that to get the U.S. fleet to 35 miles per gallon will cost $1.50 per gallon. Now is that cost-effective? That depends on what you think a gallon really should cost. Even if it is cost-effective, people aren't going to buy it. So you have a public policy problem that most of my colleagues would call, if you will, the friction, the cost of making the system somehow do it.

For airplanes, what is the cost of going from a Boeing 707 with four engines and four pilots to two engines and two pilots in a 757? That is a negative cost. And the 767 has already outdone itself three times. The airbus 340 is about to come on-line. I flew an MD-11 two months ago, an incredible plane. There it is a question more of time and that is a very important element.

To get to my vigorous intensity improvements by 2010 is going to cost a lot, although I think it is cost-effective, if energy were worth about 50% to 100% more than we pay for it today. The essential element of what we are paying for is time. We are buying acceleration and that may still pay off, but that is the difference between something you have to go out and buy and add on, versus waiting ten years until Boeing says it is coming anyway. I think that is the most important conceptual idea to keep in mind.

QUESTION: Other countries in the United Nations negotiations are asking the U.S. to agree to stabilize CO_2 emissions by the year 2000 at 1990 levels. How expensive might that be?

FRIEDMAN: Certainly Mr. Richels could answer that. I guess my answer would be, I don't know. Again, when we do our cost analyses, we get this very large range, but I would like to comment on two aspects of that question.

The first is the year 2000. The speed with which we stabilize will be a very important determinant of costs, as Dr. Schipper just pointed out. Also there are some questions of political feasibility. I started doing my first analyses of acid rain control bills in 1982. We now have an acid rain control program that will take effect in the year 2000, except for the exemptions and they will begin in the year 2004. When I say to myself, here we are almost in 1992, the year 2000 just does not seem to be a productive date to set as a target. This is obviously a very personal opinion. If we were to push that, the sorts of movements that we would have to make in that time frame, I think, could be very expensive.

The second part is that what it will cost will really depend on what policies we adopt. That is why I urge that, in our discussions on setting a target or a timetable, we must also add in some idea of the policy mechanisms that we are willing to use and the sectors that we are willing to go after. If we make transportation sacred, or if we make electric utilities sacred, if we exempt certain areas, the costs will be much higher.

SCHIPPER: I have two brief comments. First of all, whether or not you can still count CFCs really affects the problem. I personally don't think we should count them anymore.

Carbon dioxide is a very long-term problem, but that may then argue for stretching out the time frame. I think that Bob Friedman's comments are right on. It is too late. The year 2000 has been committed. Most of the automobiles that will dominate the stock by then have already been designed and are almost being made. All the power plants for the year 2000 were made ten years ago. Boeing will just be getting the 777 into the main line of production and may just be beginning to get the prop fan, which is even more efficient, on-line.

Again, part of it is deciding if it is really that bad. Even if it is really that bad, the political process is long and drawn out. I guarantee that, in Rio next year, every social, political and technical agenda, many of which are legitimate, will be advanced. We can't make the year 2000 deadline because the only way to stabilize emissions at any time is to have five-and-a-half billion individual decisions made in the world on a daily basis, not simply a couple of bills passed and a couple of standards passed.

Let me say one final word. The Danes backed off from a commitment. Now the buzz word is "pledge and review." We are doing contract work for the Danish government on their energy efficiency. How are we going to get these people to do the retrofitting? It makes sense, it is cost-effective, but they do not care. The old people won't even let us come into their homes and give them free audits. It is a long way, as Rich points out, from potential to achievement.

And you don't reduce greenhouse gas from potential. You do it from achievement. It doesn't mean that the cost calculations are wrong, but it means that, if you don't know how the human component is going to respond and you don't have a lot of time, you may miss.

And so with 2000, it is past already. It is too late. The year 2010 is beginning to be interesting. With even deeper cuts, 2025 might be even more realistic. But, again, the idea of a 35-year policy that takes away now and gives back in 35 years—few presidents have survived even two terms unscathed. Everything unfortunately argues against accomplishing a year 2000 goal.

RICHELS: Reducing emissions in the near term may turn out to be a lot easier than stabilizing—and I should use the word stabilizing—emissions at current levels over the longer term. When we talk about stabilizing emissions at the 1990 level, we are talking about putting an absolute cap on emissions. We have to remember that the economy is going to continue to grow, it is going to continue to use more energy and, over time, there will be increasing pressure upon that carbon constraint.

Secondly, with regard to the short term, we do have an abundance of natural gas. There are a number of inefficiencies within the economy that will ultimately be shaken out but it is going to become more and more difficult over time.

So to a certain extent, 2000 is a very difficult target because of the rigidities in the economy. However, don't think that the longer term is going to be any easier.

QUESTION: What must be done to improve the accuracy of the future economic analyses of global climate change?

RICHELS: I would argue that the source of disagreement among the economic analyses has more to do with one's view of the future character of the energy system in terms of supply-side and demand-side technologies and less to do with the methodologies themselves. Again, when the methodologies are standardized against a common set of assumptions, they essentially tell the same story, so I am not so sure whether the issue is one of methodological development or identifying what the critical uncertainties are and moving forward and reducing those uncertainties.

SCHIPPER: You also have the problem that I tried to allude to which is that there is so little really hard evidence that says a program here will save energy over there. So I can't turn to Mr. Richels and give him what he asked for today. I can give potential, but again that isn't what matters. I can't tell him what a DSM program will really deliver.

My colleague, Steve Nadel, says DSM saved one percent of the peak load electricity last year. One percent is not detectable in the noise. It is going to be five years before we can really see changes that are bigger than year-to-year fluctuations in the noise and that is only one small part of the problem. As a result, we are going to have to proceed on best faith. And the other thing is, how many past economic forecasts were really right? Was the present recession really forecast?

Integrating Energy, Economy, and Environmental Concerns

"NO REGRETS" TAX REFORM

Alex Cristofaro

Director
Air and Energy Policy
U.S. Environmental Protection Agency
401 M Street, SW
Room 3220M, PM 221
Washington, DC 20460

I spoke here a year ago on a comprehensive approach to greenhouse gas warming and I presented estimates of what our emissions are today and what they are likely to be ten years from now. I would like to discuss two things with you today.

First, I would like to give you an update on that work. I will then introduce the concept of taxation as a mean of reducing greenhouse gases. Specifically, I intend to discuss the possible role of carbon taxes in reducing global warming.

There are gases that are, themselves, greenhouse gases and other gases that affect the concentrations of greenhouse gases. If one is interested in knowing what the total U.S. contribution is to global warming, one has to take into account a wide number of gases. Obviously, the most important ones are carbon dioxide, methane, chlorofluorocarbons and nitrous oxide. But other gases that form greenhouse gases affect their concentrations such as volatile organic compounds, NO_x and carbon monoxide.

Each one of these gases has a different lifetime in the atmosphere and a different potential to trap heat. All gases can be indexed to carbon dioxide which is given a score of one. Methane has a shorter lifetime relative to carbon dioxide, but is much more powerful in trapping heat. Nitrous oxide also has a very long lifetime and is also very powerful in its ability to trap heat.

The scientific community has summed the heat trapping effects of individual gases over different time horizons. I am going to address the relative role of gases over a 100-year time horizon.

Once one has such an index, it is possible to put together a budget of the total radiative forcing or heat trapping ability of the nation. For the United States, if we use the indices developed by the Intergovernmental Panel on Climate Change last year, we can see the combined effects of our emissions of carbon dioxide, methane and CFCs. The U.S. total contribution is likely to go down somewhat in the next ten years because of reductions in CFCs and bounce up to around its 1990 level in the year 2010. It is noteworthy that our carbon dioxide emissions are increasing over this time horizon while our CFC emissions are going down and providing an offset.

That was a great news story, but there is now new information and people are questioning the warming indices for different gases. There are questions in particular about the indirect effects of gases. "Indirect effects" refers to the phenomenon of one gas affecting the concentrations of other gases—sometimes in positive, sometimes in negative ways. There are a lot of questions right now about the magnitude or the net effects of some of the gases.

If you look at methane, we used to think that its direct effect had a weight of about 6, and its indirect effect had a weight of 15, for a total of 21. Right now the scientific community is revisiting the indirect effects and that will determine what we think of the overall weight to be accorded to methane.

More important, though, are the indirect effects of CFCs. The story with CFCs is that, while they themselves are extremely powerful in terms of their ability to trap heat, they also deplete ozone and, because of that, there is a cooling phenomena in the lower stratosphere which may totally offset the direct effect of CFCs. Research is necessary in order to pin this down so that we will know what the overall net effect is. It could be that the decrease in CFCs that I noted previously may not be sufficient to offset the increase in CO_2, in terms of preventing our overall contribution to warming from increasing.

So what does this mean? The bottom line is that, if you just look at the direct effects of all of the U.S. gases, they are likely to increase. The big question is, what are the indirect effects? How that comes out will determine whether or not there is a net increase in the U.S. greenhouse gas budget.

Let me turn to carbon taxes as a means to reduce greenhouse gas concentrations. How do carbon taxes work? You would apply the tax to fossil fuels at the point of production and the tax would be in proportion to the carbon content. That means that coal would be taxed the most, oil would be next, and natural gas would be taxed the least. The basic idea is that, if you have to pay more for something, you will use less of it, and that is a pretty sound theory, I think.

What are the desirable features of a carbon tax? They are very broad-based and, if you think of all the combustion activities in the United States, it is very difficult to regulate each and every activity. If you place a tax on a primary fuel, then that price increase will work its way throughout the economy and all of those activities that result in carbon emissions will be discouraged. There will be a disincentive to engage in them as much as before. It is efficient because you are directly taxing what it is that you are concerned about, which is carbon, and it is very easy to administer because you just have to levy the tax at the point where the fossil fuel is produced.

We have, in conjunction with our policy research program at EPA, participated in the Stanford Energy Modeling Forum. The work involves a number of contractors and researchers. I would like to share a few insights from our work in the area of carbon taxes.

The most significant finding is that the effect of a tax on the economy and what it does to our industrial growth depend very greatly on what is done with the revenues. In fact, the effect of the revenue recycling program can be much more important on the economy than the actual energy tax itself.

What do I mean? If you are going to impose a carbon tax—and in this case the tax that we are analyzing is a \$15-a-ton carbon tax escalating at the rate of five percent per year, so that in the year 2010 it amounts to $38 a ton—that would generate some $80 billion. The overall effect on economic growth greatly depends on what is done with that $80 billion. And you could do a lot with it.

First of all, you could retire the federal deficit. What happens generally is that, when you impose a carbon tax, you raise the price of energy which, in turn, causes people to demand higher wages because they see a surge of inflation in the economy. The Federal Reserve responds by restricting the money supply in order to stop a surge of inflation. This curtails growth. The effect of a carbon tax with revenue recycled in this way is essentially a decrease in overall GNP in the early years, but because you are retiring the federal deficit in the later years, you will have lower interest rates which will stimulate investment and, therefore, the initial decrease in GNP in the longer run is somewhat neutralized.

What would happen if, instead of retiring the deficit, we use the money to cut income taxes? Essentially, you have the same sort of effect in the early years, but because you are not retiring the deficit, then the effect on interest rates will not materialize. Instead, when income taxes are cut, consumption increases. Rather than totally returning to the original level of GNP as in the earlier case, you do not get any kind of an increase over baseline GNP levels at any time.

The third scenario that we modeled was to take this $80 billion and institute a growth-enhancing fiscal policy. Specifically, we are looking at a system of investment tax credits. We are imposing the carbon tax that I described; we are taking the $80 billion and using it to stimulate investment in the United States through an investment tax credit.

We had two models where investment was not stimulated much by this, but we had two other models in which it was. In the near term, there is no inflationary surge because reductions in other business expenses are offsetting the increase in business expense associated with the carbon tax. Therefore, there is no need for the feds to take compensating monetary measures. In the longer run, the tax credit itself is going to result in increased economic growth.

What does this mean? It means that the conventional wisdom on carbon taxes is wrong. It is usually believed that they can only be bad for the economy, but when you are talking about carbon taxes of the size that are necessary to have an effect on emissions, they raise a lot of money. What happens to that money determines the overall effect on the economy and it can be positive or negative, depending on what the compensating fiscal measures are.

MULTIPLE BENEFIT ENVIRONMENTAL POLICIES

Daniel Lashof

Senior Project Scientist
Natural Resources Defense Council
1350 New York Avenue, NW
Washington, DC 20005

There is a running debate in environmental policy circles between regulatory approaches and market-based approaches such as carbon taxes. This debate is resolved here by advocating both, in judicious combination, as the most effective way to deal with the global warming problem and get a number of additional benefits at the same time.

This paper is based on a recent study that the Natural Resources Defense Council conducted in conjunction with the Alliance to Save Energy, the American Council for an Energy Efficient Economy and the Union of Concerned Scientists, in consultation with the Tellus Institute. I will first discuss policy proposals, and then the benefits that would accrue from those policies. The policies are grouped into three categories. The first is harnessing market forces, the second is making efficiency the standard, and the third is investing in the future.

By harnessing market forces, we mean a number of different measures which involve changing the relative prices seen by consumers in a way that helps achieve environmental goals. Doing this ensures that environmental impacts are reflected, at least partially, in decisions made by consumers and industry.

The first mechanism is reforming utility regulation, largely a state-level activity, and here the key item is to ensure that investments by public utilities in energy efficiency are at least as profitable as investments in energy supply. When this is done the results are quite dramatic. In California, for example, shortly after these kinds of reforms were adopted, the Pacific Gas and Electric Company announced that it had no plans to build any additional fossil fuel or nuclear capacity. It will meet 75 percent of its expected growth in energy service demand over the next decade through efficiency and the remainder through renewable energy supplies. Its southern partner, the Southern California Edison Company, the largest electric utility in the country, announced that it would cut carbon dioxide emissions 20 percent over the next twenty years. More recently the New England Electric System announced it would do even better: a 20-percent CO_2 cut by the year 2000. So utility regulatory

Global Climate Change, Edited by J.C. White
Plenum Press, New York, 1992

reform is a very powerful and important mechanism. It is important to recognize that roughly 50 percent of the carbon dioxide emissions in the United States are generated by fuel that flows through regulated utilities, either electric utilities or natural gas utilities.

The second point is to establish renewable energy production tax credits. This would be a tax credit on the order of two-and-a-half cents per kilowatt hour of electricity produced from renewable energy sources and would be available only for a limited period of time. The justifications for this approach are two-fold. First, this is an emerging industry that, in scaling up its production, clearly has the potential to become much more cost-effective. This kind of stimulus, which is based not on how much money is invested but on how much energy is delivered, can really help move the industry forward. The reason for the incentive is to help equalize the tax treatment between renewable technologies and fuel-based electricity generation technologies. A fuel-based utility can expense its fuel costs, but in a renewable system all the costs go into capital. The proposed production incentive can help equalize the tax treatment between the alternatives.

A third approach we call feebates and it involves a system of fees on inefficient equipment and rebates on more efficient equipment. This approach can be applied to everything from light bulbs to automobiles. For example, a proposed feebate called DRIVE+ is being considered in California. Under this proposal, inefficient and more polluting cars—those that are worse than average—would pay fees at the time of sale whereas vehicles that are better than average would receive a rebate. This system is revenue neutral. The fees fund the rebates. It has the very strong advantage of incorporating the costs which consumers would see over the life of the vehicle up-front in the first cost of the car. It avoids the need to rely on consumers to make life-cycle cost calculations in considering what vehicles or appliances to buy.

The final approach to harnessing market forces is to shift the tax burden from jobs and investments to pollution. In particular, we have postulated a carbon tax of \$25 per ton of CO_2 (which is about \$90 per ton of carbon).

The second category of policies, called making efficiency the standard, complements these market mechanisms that create a continuous pressure to do better, both in efficiency and in reducing pollution, with performance standards that set a minimum level of performance in a variety of areas.

Automobile standards, are a major priority. Increasing CAFE standards by 50% over the next decade is the single most important regulatory measure both to improve oil security and cut CO_2 emissions. Setting standards for minimum energy performance of housing and commercial buildings is also essential. In addition to standards for new construction, there should be requirements at the point of sale of buildings to retrofit them up to a standard based on cost-effective investments. We should also expand the current appliance standards, which have been extraordinarily successful, to other types of equipment such as motors, lighting, and commercial heating and ventilation equipment.

And finally, a revolving fund is needed to help the federal government invest in cost-effective technologies which they can identify now but don't have a readily available way of financing, given the way budgets are set.

The final group of categories, under the heading "invest for the future," involves getting a more equitable share of our research and investment dollars going into the most promising technologies for the next century. Energy efficiency and renewables, which now get about 15 percent of the federal energy research and development budget, should increase their share to something like two-thirds of the R&D budget over the next decade.

We must also build an efficient and flexible transportation network. Congress has taken an important step in that direction in the recently passed surface transportation act, which allows unprecedented flexibility for states to allocate funds to transit projects rather than roads and keep an equal federal cost share. This provides a more equitable basis for transportation planning throughout the country, and I think it could have a dramatic effect on transportation infrastructure over the next couple of decades.

Finally, we must expand education in efficiency and renewables. The kinds of measures that were taken in the 1950s to establish nuclear engineering departments, to get the nuclear industry off the ground, are now needed to support the efficiency and renewable expansion envisioned here. We must expand the pool of qualified people in every field, from policy analysts on efficiency to building inspectors, to ensure that structures have been built up to specifications.

From that suite of policies, what kind of benefits might we expect? First, in terms of primary energy requirements, we looked at a number of scenarios where these policies were implemented at different levels of aggressiveness.

In Figure 1, we start with a reference scenario that was based closely on the slightly modified Department of Energy's reference scenario in the national energy strategy. Then we record a market scenario where we implement these policies relatively gradually without including the tax shifting policies, which are based on incorporating environmental externalities into the cost of energy. A third, environmental, scenario ascribes dollar costs to emissions of conventional pollutants—such as sulfur dioxides, nitrogen oxides, et cetera—but not to the greenhouse gases, and finally a climate stabilization scenario in which we introduce a carbon tax and attempt to meet specific CO_2 reduction targets. Basically, in comparing the reference scenario to the climate stabilization scenario, looking out over 40 years, primary energy requirements could be about half of what they would be in the reference scenario, if these policies were implemented in an aggressive way.

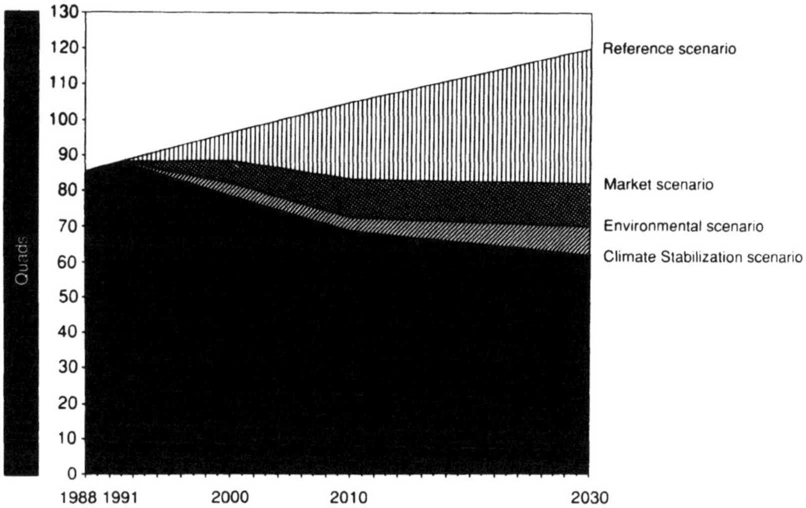

Figure 1. Primary Energy Requirements.

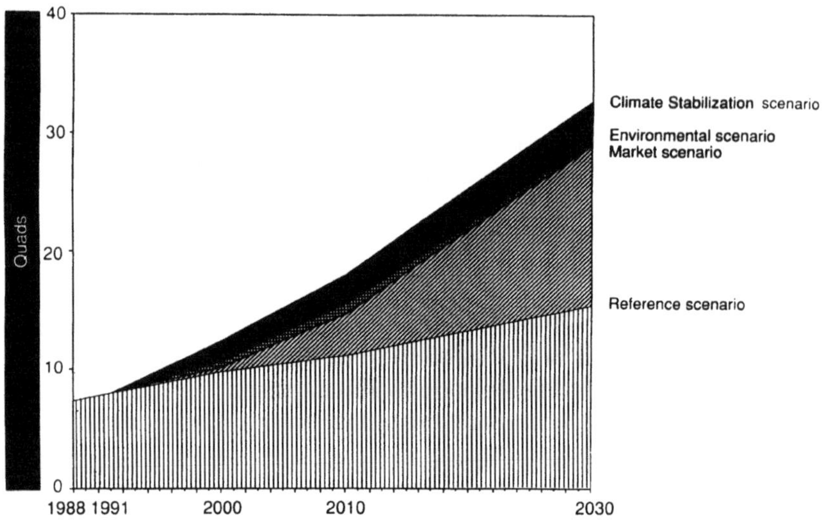

Figure 2. Primary Renewable Energy Supply.

As far as renewable energy supply, it is basically the same story. In all cases, we would expect renewable energy supply to grow, although it grows only very slightly as a share of total supply in the reference scenario (Figure 2). The total gross production of renewable energy more than doubles in the climate stabilization scenario. Because primary energy demand is cut in half, in the climate stabilization scenario, by the year 2050, half of total energy supply is supplied by renewable energy.

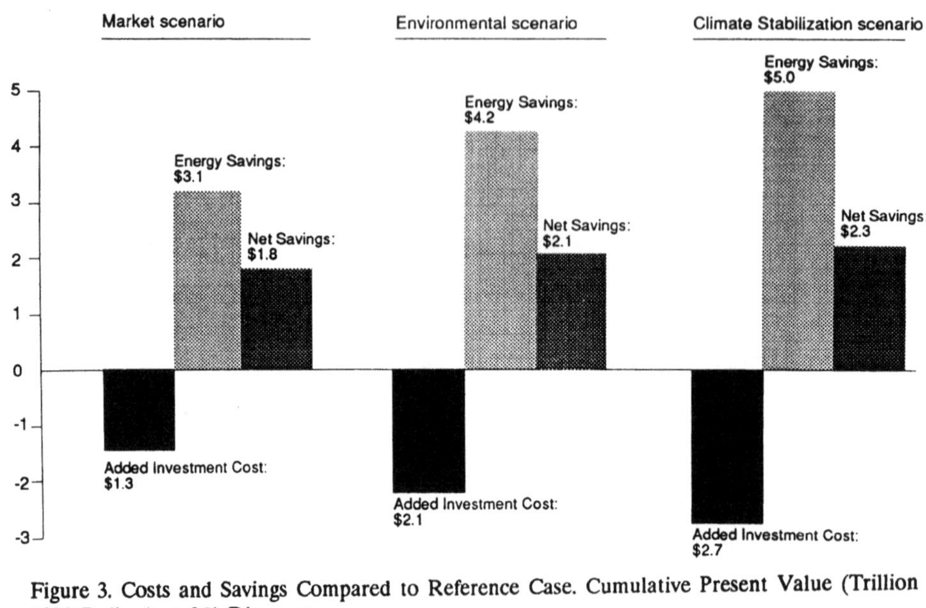

Figure 3. Costs and Savings Compared to Reference Case. Cumulative Present Value (Trillion 1990 Dollars) at 3% Discount.

These are the intermediate results. The benefits that I want to talk about are captured in the next figures. The first benefit, as shown in Figure 3, is in terms of dollars. These are good investments for the economy as well as for the environment.

In the market scenario, we are showing an additional investment cost in up-front capital to buy the more efficient equipment or the renewable energy systems as a cost, and then the gross energy savings is the second bar, shown as a benefit. That is reduced fuel consumption, reduced electricity costs, and the net is the difference between the added costs and the savings. The net present value over forty years is $1.8 trillion in the market scenario.

As we move to the climate stabilization scenario, you need to invest more money to get the benefits—$2.7 trillion, about twice as much—but you also get even more savings. The cost/benefit ratio is somewhat lower because you have to buy some more expensive equipment, but the absolute savings are larger. They rise to $2.3 trillion. At the same time, the environmental benefits are very dramatic. We were motivated particularly by looking at carbon dioxide emissions which, in the climate stabilization scenario, are reduced 25 percent by the year 2005, and over 70 percent by the year 2030. Reductions are obtained in all the scenarios, even the market scenario, but are just less aggressive.

All too often environmental policies are examined based on one pollutant that is maybe the primary one that people are looking at, but, in fact, many of these kinds of policies that emphasize energy efficiency and renewables, in contrast to some other approaches, have dramatic effects for other pollutants as well. With sulfur dioxide (Figure 4) in the reference scenario you basically meet the Clean Air Act requirements for the year 2000 and there are some additional very modest reductions beyond that, as older equipment is replaced by newer equipment meeting new source performance standards. But in the climate stabilization scenario, you not only achieve the Clean Air Act requirements in the year 2000, you get substantial reductions beyond that.

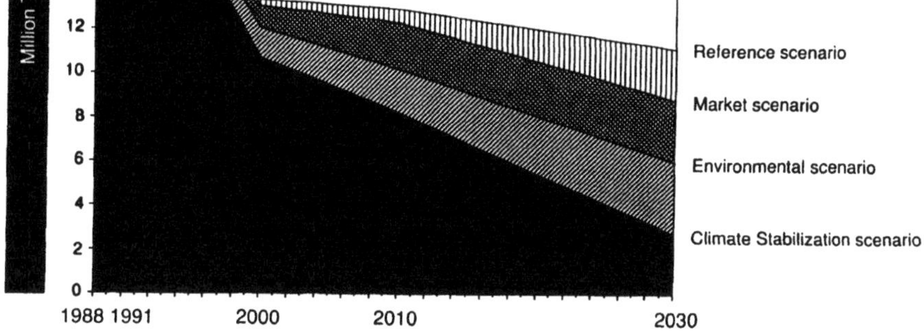

Figure 4. SO_x Emissions.

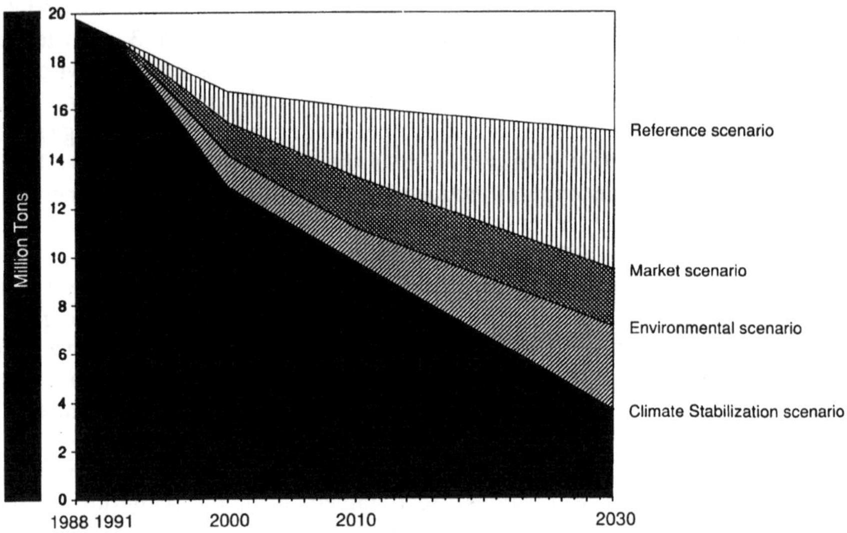

Figure 5. NO_x Emissions.

Thinking in terms of these multiple benefits is critical as companies move to implement the Clean Air Act. Some approaches to controlling SO_2 or NO_x (Figure 5) either increase CO_2 emissions slightly or leave them, more or less, unchanged. Other approaches produce significant cuts in carbon dioxide emissions as you are meeting the requirements for SO_2 or NO_x reductions under the Act (Figure 6).

One benefit that I would like to close on is the competitive aspect to the economy, because, while the United States is having a hard time making any policy decisions at the federal level with respect to global warming, the same is not the case in other countries. The European community as well as Japan are moving aggressively to implement policies to achieve the targets that they have already

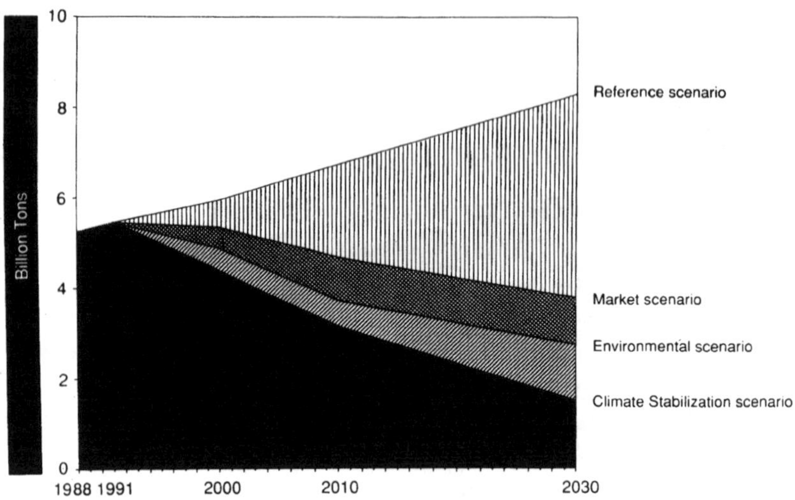

Figure 6. CO_2 Emissions.

adopted to reduce CO_2 emissions. If we do not do the same here, not only will the environment be worse off, but I think we will find our economy severely disadvantaged from a competitive point of view.

ENERGY AND ENVIRONMENTAL POLICY: THE ROLE OF MARKETS[1]

Daniel J. Dudek

Senior Economist
Environmental Defense Fund
257 Park Avenue South
New York, NY 10010

In the aftermath of both a new Clean Air Act and the Persian Gulf War, Congress and the administration have once again turned their attention to the question of the nation's use and management of energy resources. Energy, like many other valuable resources, has been the grist for many political, economic, and military battles. These struggles have largely been fought for control of these resources and the wealth that flows from such control. In our contemporary focus, we tend to forget that many of the vents of the twentieth century have been largely shaped by various conflicts over energy resources. Daniel Yergin's recent book, *The Prize*, is a compelling tale of the influence of oil. Much of the development of the western part of the United States has been spurred by energy development and use. This trend continues as the Pacific Northwest debates the environmental impact of hydroelectric development and as Alaska attempts to tap the Arctic National Wildlife Refuge.

While the drive for economic development has been fueled by both resource exploitation and strategic imperatives, energy policy over the last twenty years has been made largely by environmental regulators. Certainly there have been overzealous excesses committed by both sides. Nonetheless, this situation has evolved into a bizarre conundrum. The Clean Air Act, once the bastion of command-and-control environmental regulation, has harnessed market forces in pursuit of its acid rain objectives. The National Energy Strategy (NES) and is manifestations wending a way through Congress issue a Jericho-like trumpet call for reliance on markets. But where is the beef? When compared with the Clean Air Act, the NES looks more like a Soviet-style five-year plan than a free market manifesto. Has George Orwell's *1984* doublethink arrive a little late?

Markets and democracy, the twin pillars of our society, rest upon the foundation of free and informed choice. The conditions under which unqualified faith should be granted this system are rigorous, however, and rarely fulfilled by energy markets.

Global Climate Change, Edited by J.C. White
Plenum Press, New York, 1992

The foundation of modern resource economics is firmly neoclassical. It rests upon the fundamental notion that prices signal scarcity and can be relied upon to guide society's resource development and use over time. However, this happy state of affairs depends upon the critical underlying assumption that the markets generating these prices are operating properly. Unfortunately, in the case of energy markets this assumption is rarely true. Energy markets are notoriously plagued by externalities (i.e., economic consequences not faced by the decisionmaker). Externalities permeate the energy production and use cycle, ranging from the environmental damages of mining and drilling to the impacts of atmospheric discharges after combustion.

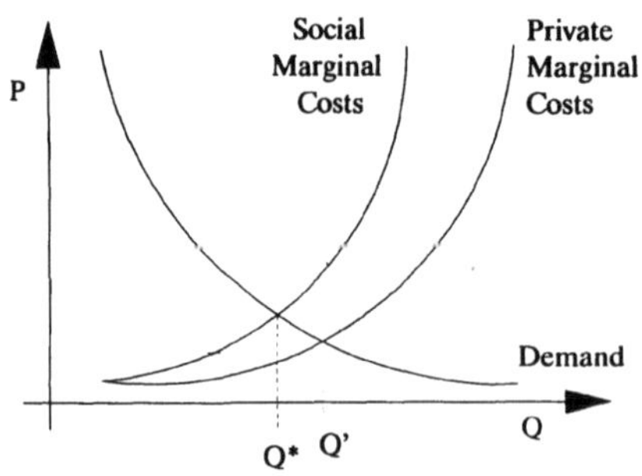

Figure 1. Externalities.

Figure 1 depicts the general policy problem posed by externalities and the implications for society. The existence of externalities drives a wedge between private and social costs. Since private decisionmakers only "see" private costs, their incentives are to oversupply energy to the market. Similarly, consumers only "see" the direct costs expressed as P prime (P') and so consume at a rate Q prime (Q'). But why shouldn't we be happy with the pleasant outcome that consumers pay less and producers supply more? The answer is that society bears the real costs of energy production and use decisions (i.e., P' plus the corresponding externality). In effect, energy producers and users are subsidized. The upshot is that we produce and consume too much energy and too many externalities.

One of the most contentious aspects of the NES proposed by the administration is the proposal to open the Arctic National Wildlife Refuge (ANWR) to exploration and exploitation. A unique wilderness area, ANWR is home to a variety of species, including the Porcupine Herd of caribou, grizzly bears, musk oxen, and nesting migratory birds. It spans and extraordinary range of environments from the harsh coastal plain to the meadows leading to the Brooks Range. The beauty of the area has been quite enough to inflame political passions; however, ANWR has also become a symbol of irrational energy policies.

After the oil crisis that struck the United States in the early 1970s, efforts shifted to develop and exploit domestic energy resources. In part, the price spike made many previous uneconomic U.S. resources recoverable and profitable. While this strategy made sense in the shortrun to meet immediate shortfalls, over the longer term, it implies that domestic reserves would be tapped and depleted irrespective of either world market prices or environmental costs. Some have questioned whether a "drain America first" policy really is prudent. Perhaps oil left in the ground would be an appreciating asset of greater value to future generations than any contemporary benefits from its combustion.

On the other hand, we have a Strategic Petroleum Reserve (SPR) that has been filled with imported oil. Why does SPR have to be filled with imported oil? Why does SPR have to be oil that has already been pumped? Why couldn't SPR provide oil resources over a range of time horizons from immediate access to long-term need? Why couldn't ANWR be treated as an addition to SPR but provide for long-term needs? The oil reserves under ANWR could be left for development under specific emergency conditions. In this way, none of the resource values provided by ANWR would be foregone. We really can have our cake and eat it too.

An accessory to our oil resource development policies is the collection of regulations focused on the automobile. The main energy policy has been the corporate average fuel economy (CAFE) requirements which mandate a minimum fleetwide average fuel efficiency. The automobile industry has complained bitterly that these regulations have been attained through downsizing with concomitant reductions in safety. Proponents have claimed large fuel savings, but analysts have been unable to distinguish between the effects of CAFE and gasoline price changes.

Environmental policies on vehicle emissions are no less a muddle. Emissions are strictly controlled at the tail pipe of new cars. Unfortunately, the slow pace of fleet turnover and uncontrolled running losses mean that the hamster will have to run faster in the future just to stay in place. Cities are unlikely to solve their urban air quality problems with new car controls. Vehicle miles traveled (VMTs) are estimated in the course of fabricating air quality compliance plans, but no coordinated policy on the relationship between VMTs and emissions has been required. Other policies, such as highway development, blatantly stimulate VMT growth.

The existence of various externalities also skews the mix of energy resources utilized in large boilers. Other things being equal, externalities distort the incentives to develop alternative energy resources. This debate has been most intense in the permitting process for new power plants. Some jurisdictions are compensating for these problems through the use of "environmental adders," cost increments assigned to each fuel in proportion to the environmental insults it generates.

REMEDIES FOR MARKET FAILURE

In the face of these market deficiencies, society could improve overall welfare by following the production and consumption program implied by Q' where both energy and externality levels would be efficient. Under this program, private and social costs are the same so that there is no divergence between private incentives and publicly desired outcomes. Both energy production and externality production would be reduced as a result of the higher cost of energy, which now incorporates external costs.

Presumably the objective of an economy is to enhance human well-being as much as possible. If human well-being is measured in both market- and nonmarket-valued components, then simplistically we can think of human well-being as composed of both normal goods and services and environmental quality. If the overproduction of goods and services produces externalities that reduce environmental quality, then at the margin, welfare could be improved by reducing the production of goods and services, reducing the level of the externalities, and increasing environmental quality. The question naturally arises as to how this market failure can be corrected so that the substantial benefits of a free market and decentralized decisionmaking can be obtained.

One of the most traditional mechanisms for internalization (i.e., making polluters face the consequences of their actions) is for the injured party to seek relief from the damage in the courts. In fact, many resource economists view the fundamental problem of pollution as a lack of definition of property rights. While the courts have helped to provide some clarity on property rights, they have more commonly been used to compensate victims of various environmental damages. Perhaps the most publicized case is that involving the oil spill from the *Exxon Valdez* in Prince William Sound, Alaska. This case will be notable for the legal precedents created in compensation for natural resource damages (i.e., reductions in the quality or size of ecosystem values).

The damages awarded to successful litigants become an economic measure of the value of the impaired environmental resource. Cumbersome as this process is, it can create incentives for firms to manage their pollution. Regardless of whether separate environmental regulatory authorities exist, firms will have some interest to control pollution if not solely to protect themselves from financially crippling pollution episodes and the resulting shock waves through the investment community.

Another response to the existence of externalities has been to create administrative processes with judicial powers to force consideration of this class of problems. Actions of state utility commissions and of the Federal Energy Regulatory Commission (FERC) provide some examples of this strategy. Most typically, this process has required review of alternative investments and operating decisions and as complete a disclosure of costs as possible. Increasingly, environmental costs have come to dominate regulators' concerns.

Perhaps the most commonly recommended internalization policy is the use of taxes or fees upon emissions discharges. The basic idea is that the divergence between private and social costs illustrated in Figure 1 can be bridged through imposition of a tax or fee. Emission fees have been most widely used in Europe. However, their implementation has diverged substantially from the recipe shown in Figure 2. Because taxes are such a politically visible instrument, the feasible levels actually levied have been quite low in relation to the marginal costs of control. In theory, the appropriate level is E* where additional control costs and additional damages are exactly balanced. In practice the result has been a significant increase in revenue collection but little incentive to emitters to alter discharges. In some

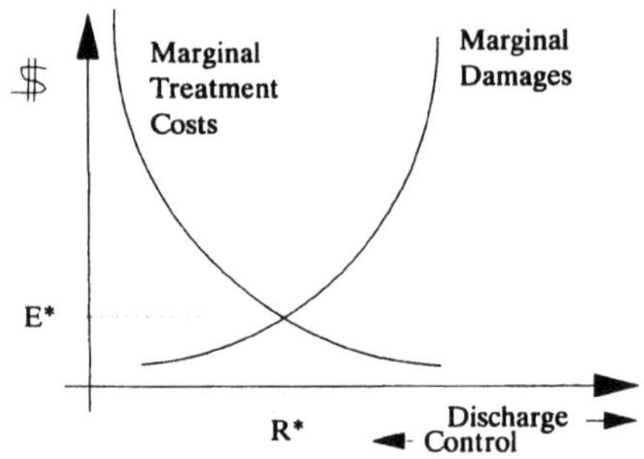

Figure 2. Optimal Pollution Levels.

cases, these revenues have been pledged to new environmental investments, but in the most perverse extremes, the funds are simply returned to the enterprises from which they were levied minus the managing agency's overhead. Often, there is not even a requirement to spend the money on pollution control. Elsewhere, governments use these revenues to pick technology winners and losers, bestowing the tax receipts on favored industries.

A relatively new tool in the regulators' toolkit is marketable permits or tradable discharge permits. These permits can be considered a type of environmental currency, an asset created from the obligation to reduce emissions, differences in control costs, and limited environmental disposal capacity. In the context of Figure 2, the environmental regulator would limit total discharges to R*. This permissible level of discharges is then distributed to polluters by some formula (gratis or sold at auction or some combination of the two). Firms are then responsible for either reducing emissions to the level permitted by their holdings of allowable discharges or by purchasing additional permits from the market. As part of its effort to strengthen the Clean Air Act, the Environmental Defense Fund developed a plan to end acid rain by employing, for the first time, a large-scale system of trading in emissions reductions.

ACID RAIN AND THE NEW MARKET IN EMISSIONS ALLOWANCES

For more than a decade, efforts to reduce acid rain through a new Clean Air Act had been stalled by powerful special interests and political infighting. Despite a widely acknowledged consensus that acid rain was damaging lakes, streams, forests and public health, environmentalists, utility interests, Congress and the executive branch remained deadlocked on the issue. The logjam was broken by the creation of a market in SO_2 allowances.

The Environmental Protection Agency issues each utility a select number of "allowances" to emit pollution. The total number of allowances is limited so that total utility emissions will be reduced by 10 million tons to half of current levels. Since the plan does not mandate a particular pollution-control technique, the first thing individual utilities are likely to do is fine-tune all their operations with an eye toward gaining extra allowances. If utilities cannot meet their pollution control obligations on their own (due to costs, etc.), they can purchase allowances from other plants that exceeded their required emission reductions. Hence, plants with maximum reductions in emissions are eligible for direct economic rewards. Because the plan rewards all emissions reductions, whether they are achieved by scrubbers or otherwise, the "playing field" between alternative energy sources and traditional sources is leveled, *exactly what the National Energy Strategy should do.*

Another benefit of emissions trading is that utilities, in seeking the least-cost method of pollution control, will likely increase their investments in conservation, cogeneration and relatively clean fuels like natural gas. The trading aspect of the plan also serves as an important platform for the development of other transboundary environmental problems like global warming. In fact, according to the EPA, the increased utility investments mentioned above will reduce emissions of carbon dioxide (a potent greenhouse gas) by 1.7 billion tons over the next decade.

Finally, to open a new plant—even to meet growing demand for electricity—a utility will be required to reduce pollution at its existing plants or pay other utilities to reduce emissions at their plants to "offset" the emissions the new plant will create. This offset requirement for new plants ensures that no matter how many new plants are needed, total emissions will never increase.

USING MARKETS TO REDUCE GREENHOUSE-GAS EMISSIONS

The major environmental problem looming over the energy horizon is climate change due to the "greenhouse effect." The vast bulk of our energy economy is based upon fossil fuels that produce as a combustion by-product, CO_2, a greenhouse gas. Traditional regulatory strategies involving prescriptive technologies or restrictions on fuels are likely to be extremely intrusive and expensive. While there is active debate over the appropriate policy response, few disagree that policy costs would be lowered if the set of available response options were increased.

Technical innovations in energy efficiency and emissions control are our best hope for controlling the degree of climatic change without seriously retarding national economies. Economic rewards of the type provided by emissions trading are the most effective mechanism to elicit the required innovations. In effect, taking relatively inexpensive and efficient action now to reduce greenhouse gas emissions can be thought of as a prudent form of insurance against ecological disaster. One of the ways to purchase the required insurance is through an offset policy. Basically, specified emissions sources would be required to compensate for their emissions by securing offsets (i.e., emission reductions wherever economically feasible). By focusing on new plants, substantially modified plants, or old facilities, the policy would gradually establish a CO_2 trading system. This policy has been recently introduced in Congress at the "Carbon Dioxide Offset Policy Efficiency Act of 1991" (COPE) by Congressmen Cooper and Synar, and Senators Lieberman and Chafee.

The legislation would require new stationary sources that had the potential to emit at least 100,000 tons of CO_2 annually to compensate for their CO_2 emissions by securing offsetting CO_2 reductions. For example, a new gas-fired 20 megawatt turbine could generate roughly 100,000 CO_2 tons per year and would be required to compensate for its CO_2 emissions. In addition, sources repowered after January 1, 1995, would be considered new sources for the purposes of this program and subject to the CO_2 offset requirement. Existing sources that reach sixty-five years of age will also be treated as new sources for the purposes of the program.

The bill establishes a number of alternatives for the production of CO_2 credits that could be used to satisfy the CO_2 offset requirement. Essentially, the bill credits emission reductions that would not have otherwise occurred. It also takes a comprehensive approach in that reductions in other potent greenhouse gases can be converted into CO_2 equivalent credits. By adopting this strategy, other important environmental priorities such as protecting the ozone layer can also be served. Examples of CO_2 credit supplies include fuel switches to less CO_2-intensive fuels, planting trees, appliance efficiency improvements, energy efficient investments, cogeneration, improved automobile fuel efficiency beyond CAFE standards, methane recovery, and the destruction of CFCs.

COPE would create market incentives for a wide variety of factors to reduce CO_2 emissions and receive credit for those reductions. These credits could be sold or transferred to new plants that will be required to compensate for their new CO_2 emissions by acquiring offset credits. This program would create powerful new incentives for greenhouse-gas emission reductions throughout the economy (e.g., through energy conservation and renewable energy).

At the present time, some progressive utilities have already committed themselves to programs either equivalent to or more stringent than COPE. COPE would reward these responsible firms. Applied Energy Services (AES) has the distinction of producing both the first real offset for a project and the first international offset. AES has adopted as corporate policy the offsetting of all new CO_2 emissions from their projects (e.g., by planting trees abroad). AES currently has several projects under way for which CO_2 offsets will be produced.

The history of U.S. participation in international atmospheric pollution control agreements is littered with problems of receiving credit for past unilateral emissions reductions. The Acid Rain program in the new Clean Air Act would create the opportunity for CO_2 reductions from a wide variety of cost-effective acid rain control investments *if they are made*!

One of the most obvious sources of CO_2 offset credits is switching to fuels with lower carbon intensity. As predicted by EPA, many of our oil-fired utility plants will switch to natural gas as the most cost-effective strategy for dealing with acid rain. In the process, these plants will also reduce CO_2 emissions.

Table 1. CO_2 Balance Sheet

Source of Reductions	Cumulative Amount (mm tons)	Source of Increases	Cumulative Amount (mm tons)
Resid to Gas	411.440 - 459.600	Added Scrubbing	30.480 - 124.700
Cogeneration	423.920	Sub-Bituminous Coal	10.010 - 19.200
Demand Response	128.850		
Conservation	728.620		
TOTAL	1,692.78 - 1,740.99	TOTAL	40.580 - 143.900

Overall there are a number of CO_2-reducing possibilities open to utilities in solving the acid rain problem. In addition to fuel switching, these include increased use of cogeneration, energy-efficient investments, and customer energy saving responses to rate increases. Table 1 shows the total tonnage of CO_2 reductions from certain strategies that would be cost-effective if SO_2 allowances were valued at $500 per ton and the total increases from two strategies that would occur at that price. *See* Dudek, Daniel J.; LeBlanc, Alice; and Miller, Peter, "CO_2 and SO_2: Consistent Policymaking in a Greenhouse," Environmental Defense Fund, January 1990.

The above discussion has made the case that markets are the best strategy to integrate our intertwined energy and environmental interests. The National Energy Strategy fails to take seriously its own recommendations to rely upon market forces. It is silent on the critical priority of creating markets for environmental resources. Recent developments in environmental policy, however, leave us with hope that this critical effort is in fact underway. The acid rain program of the new Clean Air Act points the way toward the effective use of market forces. Even more important to the health of the world economy is the development of these market-based strategies to manage greenhouse gases.

The recently adopted acid rain control program in the United States offers a significant opportunity to demonstrate the power of markets to both reconcile energy and environmental objectives as well as to coordinate the activities between nations. For example, U.S. and Canadian utilities regularly exchange both power and emissions. How will Canadian utilities compensate their U.S. partners for the use of valuable SO_2 allowances? The U.S. program will include an annual auction of allowances that does not limit participation. Further, Canadian firms could install control equipment on upwind U.S. sources and use the emissions reductions to compensate for power purchases. Or they could just pay higher prices for power. Any of these options achieves the objective of efficient resource use, a primary goal of energy policy.

As illustrated, emissions trading is a natural adjunct to trade in normal economic goods and services. The transboundary nature of many energy-related environmental problems demands policies that can accommodate separate national priorities, resource endowments, and cultural attitudes. Trade in environmental resources effortlessly bridges these differences by allowing the freedom of choice.

Even now negotiators are meeting to discuss cooperation on greenhouse-gas management. The United States has adopted a strategy of attempting to forge agreement on the mechanism of coordination and management before agreeing to targets for these gases. This strategy reflects the understanding that the same goal can be achieved at either high or low cost depending upon the policy mechanism chosen. The administration favors a comprehensive emissions trading approach to this problem as articulated in the U.S. policy position at the Intergovernmental Panel on Climate Change, Response Working Group III.

COPE, the new proposed CO_2 legislation in Congress, lays the foundation for exactly such a strategy by:

- creating the currency of CO_2 credits;
- establishing the regulatory apparatus to certify, issue, and track CO_2 credits;
- taking a comprehensive approach to reducing greenhouse gases and allowing tradeoffs among several greenhouse gases;
- creating the opportunity for international investments to demonstrate the cost-effectiveness of global emissions trade; and
- committing the United States to a trading strategy.

It remains to be seen whether the warring bureaucratic factions with jurisdiction over energy and the environment will come together on these serious market initiatives.

NOTE

[1]First printed as "Energy and Environmental Policy: The Role of Markets" by Daniel J. Dudek in *Natural Resources and Environment*, Vol. 6:2 (1991), copyright 1991 by the American Bar Association. Reprinted by permission of the American Bar Association.

LOCAL/REGIONAL POLICIES—ACTING GLOBALLY BY THINKING LOCALLY

Edward L. Hillsman

Global Environmental Studies Center
Oak Ridge National Laboratory
Oak Ridge, TN 37831-6206

This paper discusses some relationships between local decision making—that undertaken at a municipal, state, provincial or company level—and global climate change, and presents some preliminary results of some research we are undertaking to improve our understanding of these relationships.

We distinguish between two cases in which action at the local level might reduce net emissions of greenhouse gases. In the first, a city, state, province or firm decides that it is going to take some action for the expressed purpose of reducing emissions. We term this "thinking globally, acting locally." In the second, the firm or government takes action to solve some local problem, but does so in a way that presumably is beneficial to the global issue. We term this "thinking locally, acting globally."

Although we emphasize "thinking locally, acting globally" in this paper, we are interested in both types of local decision making for several reasons: the nature of the climate change issue and its linkages with local activities, the nature of political culture in the United States, and variations in the scale of the decision making process.

It is very difficult to specify much about climate change with precision. We understand general principles, but we are not in a position to say when, where, or how much the climate may change. The issue is likely to have a long time scale, and it is hard to say how climate change may affect any particular community. Given this kind of uncertainty, it is difficult to mobilize and sustain resources to deal with the issue. Figure 1, from the *Limits to Growth* study, illustrates the relative emphasis that people place on different kinds of issues. Most people spend most of their time down in the lower left-hand corner of the diagram, dealing with time scales of next week to next month, and coping with family, household, neighborhood, and local issues. The amount of time spent up in the upper right-hand corner, dealing with the next generation or next several generations, and with world-scale issues, is fairly small by comparison.

Global Climate Change, Edited by J.C. White
Plenum Press, New York, 1992

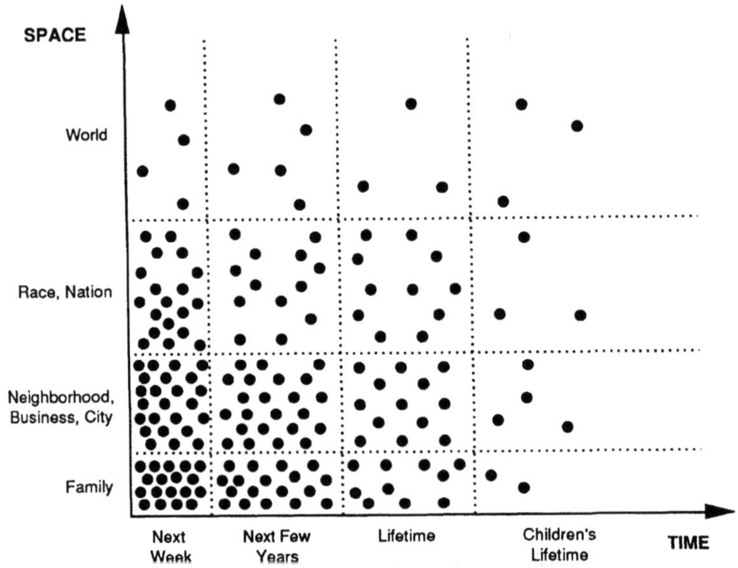

Although the perspectives of the world's people vary in space and in time, every human concern falls somewhere on the space-time graph. The majority of the world's people are concerned with matters that affect only family or friends over a short period of time. Others look farther ahead in time or over a larger area—a city or a nation. Only a very few people have a global perspective that extends far into the future. (Source: Adapted by the author from D. H. Meadows, D. L. Meadows, J. Randers, and W.W. Behrens III, *The Limits to Growth*. New York: Universe Books, 1972.)

Figure 1. Human perspectives on time and space.

However, the activities that emit greenhouse gases are largely local ones, and many cities, states or firms are beginning to recognize that these activities are also causing problems such as deterioration of local air quality, traffic congestion, resource consumption, and difficulties in disposing of waste. Measures that would be beneficial in forestalling global warming, such as improving energy end-use efficiency, often are measures that would help solve some of these problems. It may be that the global climate change issue can benefit from efforts to solve these local problems. In addition, it may be easier to motivate and sustain interest in solving local problems, where results may be easier to see, than to motivate and sustain interest in a global problem. For example, if a local government, such as the City of Washington, DC, takes measures to reduce greenhouse gas emissions, the resulting changes in emissions are unlikely to be noticed at the global scale. However, the actions that reduce emissions and the benefits of these actions may be more noticeable at the local scale.

Much of our interest has been in measures to reduce greenhouse gas emissions. However, if climate changes the changes will be noticed in local areas. Adaptations to these changes will involve local decision making. Thus, understanding local decision making is valuable from both the adaptationist as well as the preventionist or mitigationist perspectives.

It has been noted in several other papers in this proceedings that it is very difficult in the United States to achieve national consensus on energy and environmental issues. There are great differences in energy resources, other types of resources, end-use behavior, kinds of energy-using capital such as buildings, and land-use patterns between different regions of the country. On the other hand, some of these differences are much smaller within cities, states or groups of states. There is a long tradition of favoring local flexibility in solving problems, both as a way of promoting or fostering innovation—the notion of states as "laboratories of democracy"—and of permitting maximization of individual utility. Not only may it be easier for one state to act without waiting for the whole country to agree, we often believe it may be a preferable option.

In addition, there is a long tradition of resistance to what we regard as interference in local and state matters. Our political culture believes that some things are best done at the local level rather than at the national level, and we resist changes in that attitude.

All of these reasons hint at variations in the scale of the decision making process. We can see the importance of these variations if we look at transportation. Table 1 identifies five major activities in the transportation sector; the details emphasize automobile transportation, but a similar table could be constructed for other transport modes.

Table 1. Illustrative list of activities, decision makers, choices, and scale of decision for automobile transportation

ACTIVITY	DECISION MAKER	CHOICE	DECISION MAKING SCALE
INFRASTRUCTURE SUPPLY	LOCAL GOV'T STATE GOV'T FEDERAL GOV'T	capacity mode	national, local local, national
VEHICLE MANUFACTURE	AUTO MAKER	fuel economy safety performance emissions cost fuel type	national national national national, state national national, state
FUEL SUPPLY	OIL COMPANY SUPPLIER OF ALTERNATIVE FUEL	type delivery system feedstock	national state national
SERVICE SUPPLY	HOUSEHOLD MEMBERS RETAIL FIRMS GOVERNMENT AGENCIES	vehicle type delivery mode operating practice	local local local
DEMAND	HOUSEHOLD MEMBERS	trip purpose mode choice activity	local local local

All levels of government are involved in the activity of infrastructure supply. The principal choices in this activity are the capacity of the infrastructure to be supplied, and the mode—automobile, bus, urban rail, and so forth. The major decision makers are a few federal agencies, their state counterparts and local governments. Influencing decisions about infrastructure supply requires attention to all three scales.

In the activity of supplying vehicles, the key decision makers consist of only a few large firms. We have a national automobile market, and these firms operate at a national scale to make decisions about fuel economy, safety, performance, fuel type, and so forth. Influencing decisions about the types of vehicles supplied is probably best done at the national level.

The activity of service supply determines what use is made of the vehicles. Household members, retail firms and government agencies decide what type of vehicle they are going to use to supply transportation services to themselves or to their customers. These decision makers also decide what they will use the vehicle for and how they will operate it. All of these are essentially local decisions. They take place within a national culture and can be influenced by national policies, but fundamentally the decisions about whether to buy a car, whether it is a van or a sports car or a car that gets 50 miles per gallon, and how fast to drive it, are local decisions, not national ones.

Similarly, demand for transportation results from local decisions. The purposes of trips—whether a parent takes kids to a soccer game or goes to the grocery store—and the choice of modes to make trips are decided by individuals or households and fundamentally are local in scale.

Given these reasons for considering local decision making, why has so much emphasis been placed on national decision making? We think there are several reasons.

First, businesses, states and communities have concerns about acting alone. Many of these decision makers fear that if they act alone they will incur costs or other burdens that their competitors will not incur, and that they will be at a disadvantage as a result. These fears may or may not be valid, but people are basing some of their decisions on them. Action at the national level establishes a common set of requirements for all firms, states and cities.

Second, we still have a preference for the technical fix. Relying on technical fixes fits particularly well when there are national markets and a small number of decision makers to influence. It also fits with a political culture that doesn't like the idea of changing behavior for the expressed purpose of changing people's behavior. As a society we do a lot of things that change people, but we tend to refer to these as measures intended to promote growth or promote industry rather than as measures designed to change people's behavior.

Third, some events in the last thirty years have had a strong local component but required national action to become effective. The civil rights movement and the environmental protection movement in particular fit this pattern. Given the difficulties in developing a national consensus for change, some of the emphasis may be shifting back to states and localities. However, there is still a perception that, in order to solve some problems, national action is what matters.

Fourth, the abilities of state and local governments to address the global climate change issue are limited in two ways. As noted above, there are indeed decisions made at the national scale that state and local governments would find it difficult to influence. In addition, technical resources are required to implement and monitor some decisions. In some cases the pool of skilled personnel is too small for every state or major city to have its own staff, although this is not necessarily a long-term problem. In other cases it can be less expensive to have one large agency or facility rather than several smaller ones, although its larger size and complexity may make a large agency less responsive to local or state concerns.

Finally, there is much interest in having an international agreement on climate change. National governments will be involved in reaching any such agreement, and national actions will be required to comply with any such agreement.

We think it is important to understand local decision making as it relates to global climate change, and we have proposed research to improve this understanding. We proposed to do four case studies. Two of these are cases of thinking locally and acting globally; these include efforts by the Los Angeles area in California and the metropolitan areas of Phoenix and Tucson, Arizona, to improve local air quality, and by the city of Hannover in Germany to reduce energy consumption. The other two are cases of thinking globally and acting locally; these include plans by the City of Toronto and a state in Australia (yet to be chosen) to reduce greenhouse gas emissions specifically for the purpose of dealing with global warming. We have received funding for the Los Angeles/Arizona and Toronto cases, and have completed the field interviews in the former; the others are contingent upon additional funding. The objectives of the research are to improve our understanding of the following.

We want to understand how problems emerge, how discussions of these problems get onto the political agenda, and how the problems become considered potentially solvable. For example, five years ago there was relatively little interest in recycling solid waste. Now recycling is widespread in the U.S. and is becoming more so as states begin to require municipalities to reduce the quantity of solid waste going to landfills. We want to understand how changes of this kind and magnitude happen, while focusing on measures related more directly to climate change. We also want to understand how policies get formulated and adopted, and whether new institutions get created to implement them or whether governments use existing institutions for solving problems. We also want to identify obstacles to implementing policies at the state and local level, and to determine whether these obstacles are inherent in local decision making or whether there are actions that larger levels of government can take to remove these obstacles or promote local problem solving.

We also are interested in how decision makers will be defining success, how they will measure it, and what happens if the policy is being implemented against a background of rapid local growth. For example, if you make major technical improvements in energy efficiency or emission rates, and if you continue to have rapid growth, the best you might be able to do is keep a problem from getting worse. If improvement in the problem is necessary to gain political support for the policy, how do you maintain the support?

We also want to look at the role of information in the process. Does any research need to be done to help inform local decision making? Can the federal government promote local problem solving by making additional information available to communities or by presenting it differently than it does now? We have identified several needs for research and information from the first case study.

Finally, we want to know what other options exist for the federal government to promote local decision making related to climate change. In particular, are there impediments to local initiative that the federal government can remove or help overcome? Would standards be useful for evaluating success or effectiveness or costs of local initiatives?

Our present research will focus on the questions above. However, there are other questions that need to be answered but that are beyond the scope of the present study.

First, what is the potential impact of local decision making on emissions, based on the kinds of actions now being taken rather than what might be technically feasible? Some jurisdictions tend to be more active in energy and environmental matters than others; do these account for a large share or a small share of emissions? Some large jurisdictions, such as the State of California, may have some leverage over national decision making, but the requirements for and limitations of such leverage need to be analyzed.

Second, what, if any, need is there to inform local choices? We think there may be some obvious desires to resolve uncertainty about global warming, or to resolve conflicting information about the efficacy of different measures that can be taken to reduce emissions. Less obvious is the possibility that some solutions may be excellent both locally and globally if only a few states or cities pursue them, but that they may be less desirable if many adopt them. For example, competition for scarce resources could increase the cost of an option. In these cases, what kind of balance can be struck between coordinating the choices of different communities, so that not all select the same option, and interfering in their decision making? Also, when does local action become inefficient, relative to national action directed at the same problem?

Finally, we want to examine the implications of a national strategy based on promoting local initiative. Might states and local governments tackle the easier parts of the issue and leave the more difficult ones to the national government where it might be more difficult to reach consensus? If the nation were to adopt an emission reduction strategy that relies on local initiatives, what would be the implications of such a strategy for negotiating or complying with an international agreement? What is the relative effectiveness of "thinking locally, acting globally" compared to "thinking globally, acting locally"? Are the implications different for adaptive strategies as opposed to preventive ones? And, if adaptation is necessary in any event, might the resulting drain on local resources make it more difficult for states and cities to implement preventive measures?

Our first case study includes efforts by Phoenix, Tucson, and the South Coast basin in California to improve local air quality. In each instance, an environmental organization took a combination of local, state and federal agencies to court, arguing that they were not complying with the ambient air quality standards for the area. The court agreed, and compliance plans were developed.

The compliance plans in all three areas have involved a mixture of technical fixes and behavioral changes, with a greater reliance on the technical fixes. In Arizona the air quality problem was related largely to carbon monoxide emissions from motor vehicles. The compliance plans rely heavily on requiring the use of oxygenated fuels and more stringent testing of vehicle emission rates, but also include programs requiring large employers in each metropolitan area to promote reductions in commuting travel. The air quality problem in the Los Angeles or South Coast basin in California is more complex. The South Coast compliance plan relies on a more diverse mixture of technical and behavioral measures, including a similar requirement to reduce commuting travel, but the overall plan still relies very heavily on technical rather than behavioral measures. We have focused our attention in the interviews on the travel reduction measures.

Our field visits confirmed that local governments and the businesses operating in the metropolitan areas were indeed concerned about taking actions that other "competing" governments or businesses might not copy. Implementation seems to have been easiest in Tucson, where the city and county dominate the local economy. Some underlying tensions remain in the South Coast, where a balance must be struck between counties in the eastern or downwind part of the basin who receive the pollution and want stronger action to reduce it, and those in the western part where the economic activity, emissions, and most of the population are. Although decision makers in all three metropolitan areas recognize the long-term importance of land-use planning for reducing transportation demand and associated emissions, land development and ownership rights remain a very sensitive issue.

There are linkages between the behavior—commuter travel—and policy objectives, but although these linkages are theoretically sound there are some operational weaknesses. For example, agencies in all three areas think that the travel reduction programs are going to contribute only a small part of the overall emission reductions. Estimates have been made about how much the various mix of measures is likely to contribute. However, agencies are not sure how much the programs are actually accomplishing, even when they have been in place for several years. Part of the difficulty is in controlling for year-to-year differences in weather, and part in the diffuse, decentralized emission source that the automobile fleet represents.

The importance of linkage between local and global issues varies. In the South Coast area, the air quality improvement plan started purely as a case of solving a local problem. After global climate change became an issue, the air quality management district developed a strong global change component within the overall strategy. The agency did so partly to ensure that solving local problems didn't make global ones worse, and partly because it saw some opportunities to reduce releases of chlorofluorocarbons. On the other hand, among the regulated large employers, particularly those in Arizona, there is increasing skepticism about global climate change as a justification for local action. They observe revisions to the state of climate science every few months and are not sure what the current state of knowledge is. They are generally committed to the travel reduction programs for local reasons, however. Several of the firms or managers had previous experience with Southern California and wanted to avoid repetition of congestion and air quality problems in their new locations; others wanted to avoid the need for the measures as stringent as Southern California is implementing.

There appear to be several possible roles for the federal government in supporting these local initiatives. Right now, each of the three metropolitan areas is an innovator and has little experience elsewhere to draw upon. Agency staff are developing professional contacts with people doing similar programs in other cities, and a professional organization focusing on travel reduction is beginning to grow. However, several of the people we interviewed suggested that it would be useful to develop a standard design for developers of future travel reduction programs to draw upon. There is a role here for the federal government to help transfer the technology of travel reduction from innovative communities to other communities that might be interested or, under the 1990 Clean Air Act Amendments, required to implement travel reduction programs.

Commuter travel reduction programs require a lot of data, partly to support design of employer-specific programs and partly to assist in enforcement. Standards for data collection and analysis, and for program evaluation, would also be useful, as would some support for such analyses; the programs have small staffs and low funding levels and cannot afford to hire additional staff to do the analyses necessary to evaluate the real effects of the programs on travel behavior. In addition, because each program is slightly different and collects slightly different data, it is not possible to compare the results or effectiveness of the programs between cities. At least, this is what the persons interviewed in each of the three cities told us.

Finally, there was uniform agreement among government agencies and private employers that government has to be a role model. If a state, city or county requires employers to reduce commuter travel, it has to participate as an employer as well as an agency, and it needs to comply at least as effectively as it expects private employers to. Similarly, local offices of federal agencies also need to comply with the program.

What we term "packaging" is important. In each of the areas, local and state agencies had the option of developing their own compliance plans or allowing the U.S. Environmental Protection Agency (EPA) to impose one. It seems to have been very important in these three cases that the local or state agencies could point to an external threat of EPA coming in and setting standards, as a way of motivating local consensus to develop a program and then to stick with it. Local compliance with air quality standards is relatively easy to monitor, compared to emissions, and this appears to add some credibility to the threat. It is not yet clear whether a comparable mechanism can be designed to promote local consensus and action on greenhouse gas emissions.

Packaging is also important in the implementation of the travel reduction program. In some cases, businesses see the local government as regulating them. In others, businesses perceive the local government as providing them with a service to meet regulations adopted under the threat of an EPA-imposed plan. The agency in Tucson clearly wants to present itself as providing a service, a factor we think contributes to compliance by employers in the area.

A lot of small elements in program design, staffing and implementation seems to affect the success of the programs. For example, all three programs require major employers to submit plans and compliance information each year. In the South Coast, all employers are required to submit during the same calendar month; in Tucson the submission dates are staggered to reduce the peak processing load for the agency and improve its responsiveness to the employers and the information they present.

We identified several problems in coordination. Most persons interviewed commented that businesses are allowed to deduct from income on federal tax forms the cost of providing free parking for their employees, but that there are stringent limits on the amounts they can deduct for the costs of providing bus passes or other measures to reduce travel by automobile. Most people spoke in terms of wanting to provide equal tax breaks for expenses other than parking; few suggested reducing the tax break for parking instead. However, most agreed that the playing field on which travel reduction programs compete with solo commuting is not level, and that it needs to be.

A recurring problem in travel reduction programs has been that they want to get people out of their cars but cannot provide bus service or other public transportation services necessary as alternatives. The agencies that provide public transportation are not those placed in charge of travel reduction programs, although the agencies for the two activities do seem to cooperate with each other. Local initiatives to expand public transportation in the Arizona cities have been voted down. There is a lack of connection between the travel reduction objectives and the objectives of the voters who rejected the transit initiatives.

Finally, we found a lot of questions that boil down to "We don't know what the best way to deal with this problem is." As one example, consider the case of a city council member who is confronted with a proposal to develop a parcel of land as a shopping center. Given the potential for impact on traffic, congestion and emissions, is this the best use for this piece of land? If it is not the best, what is? How do I make land-use decisions for the community as a whole considering their effects during the next ten, twenty or thirty years on transportation and emissions? What options in addition to land-use are effective in reducing emissions? What options are available? What options are being researched? What is the benefit of various options, as well as the cost?

Some of these questions are very difficult to answer, and in many cases the information needed to answer them doesn't exist yet. We hope that one of the contributions of our research will be to help define a research agenda to answer them.

Although decision makers in the three urban areas are "thinking locally," there may be little global benefit from their actions to solve local problems. In the two Arizona cities, most of the improvements in local air quality come from the use of oxygenated fuels, which would have little effect on emissions of greenhouse gases unless the oxygenate is derived from biomass sources. The travel reduction measures, which should reduce emissions of carbon dioxide, are expected to yield much smaller reductions in the targeted pollutant carbon monoxide than is the use of oxygenated fuels, at least in the short run. Several of the people interviewed acknowledged this, although they believe that the real benefit of the travel reduction measures will be in the long run, by reducing the rate of growth in demand. It is too early to determine whether results will support this belief. And, as noted above, there are substantial difficulties in evaluating the real effects of these measures on emissions.

Reductions of greenhouse gas emissions seem likely to be greater in the South Coast area because the agency has made such reductions a priority, especially for chlorofluorocarbons and methane. In effect, the agency has added a "think globally, act locally" component to its "think locally, act globally" mission.

Based solely on these interviews we would conclude that the global climate change issue may benefit more from local actions directed specifically at it than from local actions directed purely at local problems—in other words, that a "think locally, act globally" strategy may be less beneficial globally than one of "thinking globally, acting locally." We would want to complete the other case studies before making this a firm conclusion.

DISCUSSION

INTEGRATING ENERGY, ECONOMY, AND ENVIRONMENTAL CONCERNS

QUESTION: For your calculations of economic gains, did you do a sensitivity analysis with regard to the discount rate?

LASHOF: Yes, we did. The calculations that we presented were based on a three-percent discount rate, a sort of long-run social costs type of discount rate. We also did a sensitivity analysis at a seven-percent discount rate and the conclusion was that you still have benefits in all cases, but smaller net benefits. They are about .06 trillion dollars, roughly equal in all three scenarios, compared to the couple of trillion dollars found at the corporate discount rate.

QUESTION: What is NRDC's position on COPE?

LASHOF: We have some concerns with the specific way that bill is structured in the stand-alone version. We don't have a predetermined view of whether a tradable permit system or a carbon tax system should be implemented first. We tend to think that a carbon tax approach might make more sense at this time. For us to support a trading system we want to ensure that there is a firm cap on the emissions that are regulated under that system so there are possibilities for crediting things which might have been done anyway against real increases in emissions. We are concerned that that might not be sufficiently guaranteed in the way that bill was written as it stands.

QUESTION: Thus far, in the Clean Air Act implementation, in Phase I of the acid rain title, we are not seeing much participation by electric utilities in the allowance trading program. Do you see a reason for confidence that the allowance trading program is eventually going to work better and therefore it may work well in a carbon dioxide or greenhouse gas context as well?

LE BLANC: Yes, I think it is really still quite early. As you know, the allowances don't really get allocated until 1995. Right now utilities are going through the process of compliance planning to decide exactly what they are going to do, and there are still a lot of regulations that have not been issued. EPA has come out with the proposed rules for the bulk of the allowance trading and the SO_2 monitoring regulations, but at the state level there are a lot of decisions by the public utility commissions that have not yet been made.

I think the market is going to pick up substantially in Phase II. In Phase I there are really only about 111 affected plants that will be trading allowances. In Phase II it will be more or less the universe of utility units.

I am very confident that the market will take off and, as you know, the Chicago Board of Trade is going to be a central exchange for sulfur dioxide allowances. They have also applied for a futures market in SO_2 allowances, so there is some confidence that something is going to happen.

QUESTION: Do you see a benefit in going with the offset approach that has been proposed instead of going with an approach more analogous to the acid rain title where you would roll back emissions with time and have a cap? Is there a particular reason for using the offset approach?

LE BLANC: I think ultimately there will be a system more analogous to the acid rain provision with a cap. The Cooper bill was intended to be a first step to establish the carbon dioxide bank and the certification mechanism, to get these in place for certifying the offset of reduction credits. This was intended to be a first step toward setting up a more comprehensive system.

QUESTION: What happens to the economy when a carbon tax is used to increase spending?

CRISTOFARO: We have not analyzed that because, basically, the budget deficit is probably large enough already.

QUESTION: What economic environmental effects would result if carbon tax revenues were recycled into environmental projects or into subsidizing energy efficiency and renewables?

CRISTOFARO: The drag on the economy from energy taxes occurs because there is an inflationary surge due to the imposition of the energy tax. If you can devise some kind of mechanism to reduce prices by offsetting business expenses in some other way, such as recycling carbon tax revenues into energy-efficient investments, then you tend to mitigate that initial inflationary surge, which then means that the federal reserve does not need to restrict the monetary supply so much, which means that the economy won't experience as much of a decline.

Obviously we didn't analyze that case in any detail, but the effect of recycling revenues to other business expenses—to offset other business expenses—would be to mitigate their impact.

QUESTION: Has the success of recycling been the impact of kids learning about it in school and influencing their parents? Can we do more through this channel of education to change behavior? How does one solve the problem that many local activists are more obstructionist than constructionist—the Hinders Project, new hydropower, mass transit, nuclear waste disposal and so forth.

HILLSMAN: Public education involving the kids and so forth has definitely been a part of it. I suspect that the real impetus for this, however, was the sudden realization by a lot of communities of the cost of building new landfills, in part because of the Not-In-My-Back-Yard (NIMBY) syndrome, and also the increasing regulations on how those landfills have to be built and operated.

In terms of what we could do on the public education end of it, I think that that is going to be more of a local decision than a federal one. We tend to resist the idea of having a federal curriculum and federal control of education, and I do not mean that in a pejorative sense, but it is the feeling that a lot of people have regarding that kind of activity. I think it is something that will be significant at the local level and it will be a local decision, most likely not a federal one. Public education might be something that nongovernmental organizations, environmental groups, want to promote, but I don't see the federal government doing that, at least not immediately.

How to solve the problem of obstructionist groups? I don't have an answer for that. Part of that goes back to this organized concern of not wanting to put yourself at a disadvantage with some kinds of facilities, like toxic waste facilities. For a long time some of these facilities were seen as beneficial, then a lot of communities found that they are built by outside agencies and couldn't really be taxed. Communities are being asked to bear the costs of installing these facilities and are not receiving a proportionate share of the benefits. That is part of the NIMBY syndrome.

QUESTION: The Data Resource Inc. (DRI) model has been criticized for failing to account for intergenerational effects and long-run impacts on the interest rate. Does it apply to the other models you use? What are the implications for your results?

CRISTOFARO: I think the DRI model as we used it essentially talks about an effect on the economy over the next twenty years. I am not sure what is meant by intergenerational impacts. If what is meant is the need to take into account the effect on the economy in the year 2050 or 2060, then it is correct that the DRI model doesn't do that, but I am not sure that any model does to anybody's satisfaction.

QUESTION: Is this a valid criticism? Does it apply to the other models you use? What are the implications for your results?

CRISTOFARO: That was part of the question on the long-run growth and part of the DRI story is that, in the deficit reduction case where we are retiring national debt, it shows that, as you lower interest rates, you will affect capital formation. You will have more growth in the economy, more investment. We had a short-trend dip but in the long run there was a gain in the economic growth over time. So the model specifically addresses the effect on long-term growth, one of the advantages of this approach. Several of the other models work in the same fashion as the DRI model does in that respect.

QUESTION: What happens to emissions of CO_2 as a result of the three ways of managing revenues from the $15 carbon tax? Were there sizable emission reductions in the third case in which economic growth increased?

CRISTOFARO: That is a very good question. Actually we presented four models and each model has a different estimate of what happens to carbon emissions in response to an energy tax. It is true that in some of the models, if you use the revenues to stimulate the economy, then that in turn results in an increase in carbon emissions. Part of the decrease that you get because you are imposing a tax is made up by a kind of income effect, because as the wealth of the economy grows, more carbon is used, and that is more important in some of the models than in others. I can't really tell you right now what the magnitude of the carbon reduction is for each of the four models, but I would be happy to provide that information if somebody were interested.

QUESTION: Could you comment on SO_2 particles in the atmosphere as having a cooling effect?

LASHOF: There is some recent research that is emphasizing the possibility that SO_2 particles in the atmosphere do exert a cooling effect which is of the same order of magnitude. That may be one reason why the earth has warmed up by half a degree centigrade over the last century rather than one or two degrees centigrade, which is the range of warming that should occur based on the greenhouse gas increases. There is still a lot of uncertainty about that, but it is receiving a lot of attention.

QUESTION: What about the reported cooling effects of SO_2? Is it true? If so, why did we legislate it out?

HAY: I don't think this is a new concept. It is a concept that has been around for a long time. I must say that sitting in on many of these discussions, I often find them very troubling because of the concept that we can fine tune what we are doing, that we have a complete knowledge and that, in fact, we ought to in fact pollute intentionally because we think it is going to have a particular effect. I have trouble with the necessary focus on mitigation as opposed to prevention. I have seen it seriously suggested that we ought to be increasing our sulfur dioxide emissions for climate purposes, even dropping sulfur dioxide from airplanes. Let me say we don't have nearly that good a fix on what it is that we are doing.

QUESTION: Would you comment on the recent finding or correlation between sunspot frequency and temperature?

LASHOF: There was a recent paper that showed a pretty strong correlation over a relatively short period of time—the 100 years for which we have records in the length of the change of the sunspot cycle and the temperature. There is no mechanism known to cause that correlation so there is certainly a strong possibility that it is mere chance.

All the evidence is that the sun's output doesn't change by more than a few tenths of a percent, both by observing the sun, observing other stars and models of the sun. If the earth's temperature is really increasing and decreasing on the order of half a degree centigrade from such small changes in radiated forcing, then there is no plausible mechanism that would prevent the earth from also responding with a similar sensitivity to the increase in greenhouse gases, which represents a much stronger overall force.

QUESTION: It is usually assumed that biomass fuel use yields no net carbon emissions, i.e., that biomass fuels are produced in a manner which is permanently sustainable. Thus, carbon taxes are usually discussed for fossil fuels but not for biomass energy. Is there not a serious risk of encouraging the nonsustainable exploitation of biomass energy, and wouldn't it make more sense to tax biomass fuels like other carbon-bearing fuels while using separate measures to encourage biomass production?

LE BLANC: I don't understand why that would encourage the nonsustainable use of biomass.

LASHOF: I think it is a legitimate possibility. If you are taxing fossil fuels and not biomass energy, that will make biomass energy more attractive, which is one of the things that you want to do to the extent that the biomass is being produced in a sustainable way.

I think you can deal with that by a tax and refund scheme where you tax biomass fuels at the point of combustion and provide some kind of rebate for the carbon sequestration values. But even doing that doesn't solve the problem because it could create an incentive for cutting down old growth forests, which is what you want to protect. So I think you need, in addition, environmental regulations to protect the ecosystem values. We care about the forests because they are reservoirs of carbon and for a lot of other reasons.

QUESTION: How large do you anticipate the administrative costs being for an emissions trading system compared to a traditional command and control system?

LE BLANC: I don't think I have numbers for climate change. For acid rain, the utility compliance costs between the current system of emissions trading and a command and control system as embodied in the Weiss bill, that at one time was introduced into Congress, were about $3 billion a year. We took those numbers and plugged them into DRI's model of the U.S. economy and I unfortunately can't remember the exact number we came up with, but it was magnified in terms of its impact on GNP.

The most interesting finding from that model run to me was that the job loss from the command and control approach exceeded the job loss from emissions trading by a factor of about ten. In view of the controversy over loss of coal mining jobs if you don't go to a command and control approach and install scrubbers, I thought that was an interesting finding.

HAY: We have talked about a spectrum of approaches from command and control through a carbon tax through tradable permits. From my parochial view in the natural gas industry, and I think from other perspectives, I tend to favor the allowance trading sort of approach. The carbon tax approach is attractive but you have more intervention there by some regulator setting the value of the carbon tax.

Still, in our industry we recognize that, as in the Clean Air Act, we are probably going to have a mix of these things to face eventually. We are going to have to do some things to reduce our methane emissions from natural gas operations, for example.

I am beginning to feel that the way these things are implemented is as important, if not more important, than the choice of the basic approach. We have seen that effect in appliance efficiency standards; by focusing on end-use efficiency, not on full-cycle efficiency, we have caused a shifting of load from gas-fired furnaces to electric furnaces, with less energy efficiency and higher emissions of greenhouse gases in many instances.

New hot water heater standards potentially might double or triple the cost of a gas hot water heater in order to increase the efficiency of that device by 10 or 20 percentage points. A gas hot water heater might be 70-percent efficient at the end use versus an electric hot water heater that is nearly 100-percent efficient, but when you take into account the full cycle including electric generation, you are looking at maybe 30-percent efficiency for the electric device and maybe 65-percent efficiency for the gas device.

We face the same thing with regard to carbon taxes. Many in the Natural Gas Association think carbon taxes might not be too bad but how you implement them is really important. If you are in an area that has, say, a mix of two-thirds nuclear generation and one-third coal generation for your electricity, and you put a $1 tax on gas and a $2 tax on coal, you end up raising the gas price that a residential user sees, say, by maybe 20 percent and the electric price by only 10 percent. If that area happens to have surplus coal-powered electricity generating capacity, the result is a switch to the coal-powered electricity and increase in the output of greenhouse gases.

QUESTION: Do you anticipate ways to structure a carbon tax, for example, to actually achieve what the target is, to avoid undesirable fuel switching and achieve desirable fuel switching in terms of greenhouse gases?

CRISTOFARO: The most efficient way to raise the tax would probably be at the point of primary extraction. That tax would work its way through the entire life cycle of the use of the energy so if you lose half the energy in some conversion process, then the effect of tax at that point is doubled and it reflects the loss of efficiency in the conversion process.

HAY: A very important factor for the gas industry is the cost of electricity as seen by the end user; as much as two-thirds of that cost may be in capital charges.

One reason that more of my members don't support a carbon tax is that they actually see it as shifting away from natural gas, and they think gas ought to be part of the climate solution.

Related to that, one problem with the carbon tax is not covered here. The affluent can afford higher energy costs; the poor cannot, so they will be subsidized.

The result of these two effects is that energy use will not decrease, or at least not as much as hoped.

LASHOF: I agree with you that a lot of these things depend more on how it is implemented than on what sort of ideological spin you put on it in terms of what instrument you use. But in terms of the specifics, I think you do need to look at the distributional impacts of the carbon tax and recycle part of the money in a way that can offset the disadvantages.

Rebating the money based on how much people pay would essentially eliminate the tax from some sets of people, and would not diminish the benefits of the tax very much. If you recycle the money in a way that increases the net income of poorer people, it is likely that most of that additional income will not go to buy more energy; it will go to other uses. If you work through the numbers, there may be a very small take-back effect but it is not that significant.

With respect to the fuel-switching point, I really wonder whether that is an artifact of the way the analysis is done based on percent changes in prices versus looking at absolute prices and at what is the cheapest price for the consumer. If it is not an artifact of the analysis, then I would suggest it is an artifact of market imperfections where consumers do not do a full life cycle cost. If you do a full life cycle cost, natural gas is to the advantage economically compared with coal-powered electricity and you would prevent fuel switching. The way to get around that is to implement various mechanisms such as feebates which help consumers respond to life cycle costs.

HAY: We have done a lot of arguing about this for a long time and our perspective is that when you are looking at, say, $6 gas and $20 electricity, it is the percentage change that drives the decision at the margin, not the absolute change.

QUESTION: Should we also have a nuclear waste tax to balance the carbon tax?

CRISTOFARO: Actually, the tax proposal which is being discussed within the European community is focusing on whether there should be one. But if there is one, they are looking at a tax that is 50 percent based on BTUs and 50 percent on carbon, simply because they are concerned about encouraging the development of additional nuclear energy if it is only a carbon tax.

QUESTION: Can you compare the various approaches in terms of carbon tax, allowance trading, command and control efficiency approaches, and some combinations?

HAY: Let me raise one more point, if I may. Dr. Lashof had a reference scenario in which both primary energy consumption and greenhouse gas emissions are rising fairly steeply. One of the new things that we are starting to do at the American Gas Association, with our standard long-term energy outlook, is to calculate the greenhouse gas implications. We are seeing in many of our scenarios that we don't have a very substantial rise in greenhouse gas emissions or, indeed, a very substantial rise in primary energy consumption.

I just wonder how current you feel your business-as-usual reference case is.

LASHOF: Our reference case is based on the Department of Energy's national energy strategy reference scenario. We made a couple of modifications to it which reduced the growth in primary energy and CO_2 emissions in the reference case compared with them, because we thought they were egregiously unrealistic, particularly the growth in energy-intensive industries that they assumed. Also they assumed continued long-term growth in vehicle miles traveled with very limited saturation effects, which we didn't think was realistic. So our reference case is a little bit lower than theirs, but we wouldn't say that we endorse that reference case as reality based on current policies. I think it is quite possible that emissions might be significantly lower than that, based on current policies.

Our focus is on the alternative cases so the market case which incorporates a number of policies, but not very aggressively, starts off with relatively flat emissions and then these decline over time. It is unclear how strong a set of policy changes you need to get there because there are a lot of tendencies that are already moving us in that direction.

QUESTION: Could each of the panelists make a short summary and perhaps add comments?

LE BLANC: I think I would like to state the reasons why we have focused on a policy of strict standards coupled with trading rather than a tax as a solution to the global warming problem.

I think the first and foremost reason is environmental uncertainty, and especially if this is in response to a climate convention where we have international obligations to meet a certain target. How do we know that a tax is going to achieve that goal? At what level do we set the tax to make sure that we reach the environmental goal that we want to reach? I really don't think anyone can answer that question. Monitoring is going to be required in any case, to monitor emissions or monitor the fuel use and to make sure that we achieve the target.

Unless the tax includes some kind of provision for rebates, either for carbon sequestration through forestry, through tree planting, as well as rebates for capture and reduction of other greenhouse gas emissions, there are going to be enormous losses in efficiency and cost-effectiveness.

So the other reason is the political feasibility. It has been very difficult to even get the few cents of gasoline tax maintained. In fact, that has been very difficult to do and that is a very small thing. When you are talking about taxing carbons at these really high levels, I think some consideration should be given to the question of political feasibility.

Over time, the underlying price of fossil fuels can change and the effect of the tax can be negated. A tax may be set at a level today that might achieve a certain result, but if the underlying price of the fossil fuel goes down for market reasons, that tax might not be effective five years from now. So you might be in a position of constantly trying to change the level of the tax.

The pluses of trading are, first of all, the environmental certainty. You set a standard, you monitor it and you make sure that that standard is achieved. It provides real incentives for polluters to control cost-effectively. It also provides polluters total flexibility in how they control and gives them the opportunities to use offsets such as tree planting and reductions in other greenhouse gases.

HILLSMAN: We have not gotten into our other case studies where we are focusing on direct attempts by communities and states to reduce greenhouse gas emissions specifically. I am more optimistic about the potential for that kind of measure to reduce greenhouse gases than I am for attempts to solve other problems. I am hoping to get large improvements in greenhouse gas emissions as a by-product, based on the case studies we have done so far.

I think that there is a real possibility of dividing up the responsibility for reductions between the federal level and the state level, the federal level setting some objectives, telling the states and cities: do what you do best to meet these targets, and we will do the other things that need to be done as well. I don't think the states and cities can do it alone. I don't think the federal government can do it alone, if we are talking about serious reductions.

LASHOF: I think our answer is, as I said at the outset, all of the above. We think that you need a combination of different kinds of measures, performance standards, tax reform, and probably tradable permits being phased in over the longer term.

There are clearly two problems that we need to address and they need to be addressed in different ways. One is that environmental costs are not incorporated in market prices. You can incorporate them in market prices either with a tax or a permit trading scheme.

The second is that, even if you have the prices right, the markets aren't working properly, the consumers are not making decisions that are based on life cycle costs. There are lots of reasons for distortion in the market. You have to address that in a number of ways, which includes efficiency standards.

As far as taxes versus permits go, we would probably say to use both, or you should option the permits, and the reason is that there is a difference between CO_2 and SO_2. The SO_2 trading scheme is in a regime of regulated utility. The rate of return is controlled by state commission. When you go to the national economy and you are controlling CO_2, we would have a real problem just giving those permits away to existing polluters. That violates what I would call a polluter-pays principle. You need some way to recover those revenues, either with the taxes as is being done with the CFCs right now with a limit on production, or by optioning the permits initially which is in many ways equivalent to a tax.

CRISTOFARO: If Dr. Lashof's position right now is all of the above, I guess our position right now is none of the above.

Essentially the first question you have to focus on is whether this is a problem that warrants a policy response at this time in the direction of expensive mitigation. But putting that aside for the moment, the choices are between taxes, permits and regulation. I think there is one important choice that was left out, which is technology cooperation with developing countries. Our emissions are likely to stay constant or to go up or down rather slightly. The real action is in developing countries, which tells me that technology cooperation could be a very important aspect of any climate change policy that is geared toward reducing emissions in the future.

Enough has been said about the other options. Obviously, with permits the advantage is the certainty of the environmental outcome but there is uncertainty with respect to the control costs, whereas with taxes you have relatively more certainty with respect to the costs but less certainty with respect to the environmental outcome. How you feel about knowing more about the total cost of the control program versus the significance of differences from emission outcomes will determine which one you prefer.

HAY: When I think about the global warming debate, I am struck by the similarities with the acid rain debate, and that was mentioned earlier by another speaker. I would like us to think back for a moment over that controversy which ended only recently with the passage of the revised Clean Air Act. Remember that this debate also involved scientific controversy and it pitted industry and environmentalists in an ever-escalating battle that took thirteen years to resolve. That debate took thirteen years and only $5 billion a year was at stake.

Now we are facing the global warming case. The financial stakes are perhaps 10 times higher—$50 billion per year—and scaling up the 13 years by 10 times, we may have to wait 130 years to resolve the global warming problem. That is if the actors in the debate take their traditional adversarial roles.

Now there is an alternative. We can learn from the mistakes that we have made with acid rain and focus on innovative options. When I am on a podium, I always talk about this. I think what we need to engage in is a politics of vision rather than solely relying on the politics of blame. Industry, environmentalists, we will fail the world if we spend decades pointing the finger of scorn at each other. But I can tell you it can be lots of fun to fight—I have made a career of it—but all too often when these powerful lobbies lock horns, we end up with emasculated legislation that burdens business with little environmental payoff.

Instead, those of us engaged in the debate need to propose solutions from the very beginning that are sensitive to the concerns of the constituencies involved, and government needs to bring those constituencies together and get them to enter into a meaningful dialogue. I have had some experiences with this, which is called negotiated conflict resolution, where the actors actually negotiate with each other. I have found that there are a lot of people out there in industry, government, and environmental groups who are, indeed, capable of going beyond the politics. They are capable of developing and shaping a vision of a world where the global warming problem is solved in a way that is politically acceptable. However, there is a catch. Before any of us can come up with a meaningful political vision, we need to understand the perspectives and the concerns of the key actors.

We will hear from a number of sectoral representatives. It is my hope that their remarks and the questions that you people prepare will lay out for us glimmerings of common ground. I charge the presenters with educating you about what each sector's key concerns and needs are in dealing with global warming. I make a similar charge to the audience in your questions. I think it would be particularly helpful if you were trying to explore options that the panelists have not considered, options that may resolve some of what they have expressed.

Implications for Decision Making

Sectoral Perspectives

IMPLICATIONS FOR DECISION MAKING: GOVERNMENT PERSPECTIVES

Howard Gruenspecht

Associate Deputy Under Secretary for Program Analysis
Office of Policy, Planning and Analysis
U.S. Department of Energy
Washington, DC 20585

I would like to focus on the implications for decision making in three areas related to policy towards greenhouse gas emissions. The first involves issues related to the apportionment of risk, which have already been briefly discussed by previous speakers. The second concerns the standard to be applied in evaluating costs. The third encompasses the international equity issues raised by policies to reduce emissions of greenhouse gases.

The issue of emission certainty versus cost certainty was raised by Alex Cristofaro. Fixing emissions maximizes cost uncertainty while adopting a fixed set of actions or taxes leads to uncertainty regarding the resulting emissions outcome. This tradeoff have has been recognized in the academic literature since the seminal article by Martin Weitzman. Several relationships are important in determining the allocation of risks that best serves societal interests. One is the link between regulated and unregulated emissions. Another is the link between total emissions and adverse impacts. Finally, there is the link between policy and the benefits of policy.

The transparency of these linkages varies widely across different regulatory contexts. The Clean Air Act is an example in which they are fairly clear. Emission standards under that Act are designed to meet national ambient air quality standards. Target air quality standards apply at all points within the United States, and attainment of these standards is virtually unaffected by foreign emissions that are not regulated. Finally, the ambient air quality standards are set to protect public health with an adequate margin of safety.

In addressing global climate change concerns, the picture is much more uncertain in all respects. First, the link between national and global emissions is uncertain. This linkage is critical because it is global, not national, emissions than can potentially affect climate processes. Several different channels of influence can be identified.

Global Climate Change, Edited by J.C. White
Plenum Press, New York, 1992

For example, if a subset of countries limit emissions, very energy-intensive industries might move to other countries raising their emissions. Or, emissions limits in a few major oil-consuming countries could lower world oil prices, allowing cash-constrained countries to increase oil consumption. These types of market responses would result in a less than one-to-one passthrough of national emissions reductions at the global level.

Alternatively, emissions limitation policies in some countries could result in a trickle-down of technologies for application elsewhere. Some sort of a trickle-down technology process is implicit in the notion that national emissions limitation policies in a small subset of countries can make a difference on a global scale.

On the link between global emissions and climate, many scientific issues need to be resolved. For example, there is the issue of offsets to the radiative forcing from CFCs via the depletion of stratospheric ozone. There are also many questions related to the baseline against which the forcing impact of an increment of emissions should be evaluated and the handling of the "missing carbon" problem in cross-gas comparisons. The role of influences other than greenhouse gases -- such as atmospheric aerosols -- on climate also merits scrutiny.

Finally, many questions arise regarding the benefits of emissions reduction. In industrialized countries, we know that the great bulk of activities are not sensitive to climate change. Developing countries have a higher proportion of output in potentially climate-sensitive activities, but it remains unclear how climate change would actually affect those activities. For instance, various studies done on the impact of climate change on agriculture show uncertainty even as to the algebraic sign of those impacts.

It is very important to recognize the key distinction between the impacts of climate change as a whole, whatever they may be, and the benefits of emission reduction policy. The benefits of emissions reduction policy depend on the *difference* that policy makes in the timing and magnitude of climate change. Given expected global emission patterns, the benefits of emission reduction policies undertaken by a limited group of countries are likely to be a small. To pick a specific example, a commitment by all the OECD countries to cut their carbon emissions to zero--which is inconceivable -- would alone be insufficient to hold global carbon emissions in 2025 at the current level. That speaks to the global dimension of this issue.

The points outlined above, which illustrate how different the climate change issue is from some of the standard air pollution issues that we have traditionally dealt with, bear on the issue of where the risks should be borne. In view of the Administration, the establishment of arbitrary targets for U.S. national emissions of individual greenhouse gases represents an inappropriate policy response. The Administration's view on how to proceed has been set out quite well by Alex Cristofaro--market reform, technology development, technology cooperation and, finally, informational and educational activities. These steps may get us a surprisingly long way.

DOE has been doing well with efforts to promote utility integrated resource planning (IRP) to promote conservation. EPA is complementing these efforts with their "Green-Lights" program, which points out opportunities to invest in energy-efficient lighting to major commercial, industrial, and government entities. Reforms in the natural gas market which allow for the increased use of natural gas, which has a lowest carbon content per unit of energy than other fossil fuels, and other measures outlined in the national energy strategy can also significantly reduce greenhouse emissions.

A related matter is the issue of how to evaluate emission reduction costs. The Department of Energy (DOE) has just released a report to the Congress called "Limiting Net Greenhouse Gas Emissions in the United States." It is a two-volume document addressing a Congressional request for DOE to analyze policy options to achieve a 20-percent reduction in carbon dioxide emissions by 2000 and a 50-percent reduction by 2010.

Let me briefly review the findings of the DOE report. We found that there is a very large cost of meeting short-run emission reduction goals. To get a 20-percent reduction by 2000, a carbon tax in the range of $500 per metric ton is needed. To stabilize emissions in the year 2000 at current levels, a carbon tax of about $140 per metric ton is needed. If the focus is placed on "net" emissions, tree-planting programs to sequester carbon have a lot of promise as an intermediate measure, running from now to about 2030. However, over the longer term land availability becomes a serious issue for tree planting on a massive scale. We found that timing is essential and that the development of more efficient technologies, low-or no-carbon technologies, really makes a difference in lowering the costs of greenhouse gas emission reduction. And, finally, while conservation is very important to any effort to limit emissions, conservation alone cannot do it all. New supply technologies must play a role.

You have heard at this meeting about differences between the "top-down" and "bottom-up" approaches to evaluating the costs of emission reduction. The U.S. government is quite eclectic in this regard, --official Washington looks at both top-down and bottom-up models. The European Community (EC) and other proponents of early emission reductions have tended to focus on the bottom-up approach. However, recent developments in the EC, notably the debate over the $10-per-barrel-of-oil tax (which is equivalent to about $84 per ton of carbon) indicate a realization that carbon emission reductions may not be quite so easy as the bottom up models suggest. Obviously, tax policy remains controversial in the European community.

In looking at how to evaluate emission reduction costs, it is also important to recognize that we need to evaluate policy based on what we will do, not on what we might do. In early 1990, the Clean Air Act debate stalled on the Senate floor for six weeks over the transfer of some coal mining jobs from the high-sulfur coal region of northern West Virginia to the low-sulfur coal region of southern West Virginia. There were actually no net West Virginia jobs at stake, only a question of where in the state they would be. And that question was so sensitive the debate was held up for six weeks.

Optimization and economic models agree that the coal industry would be hit hard by *any* efficient emissions reduction strategy. Given the history of the clean air debate, one might question whether we would really do it that way; a politically acceptable approach could well be considerably less efficient than the model solution. In this sense, the top-down models, far from overstating emission reduction costs, may fail to account for important factors in the policy process that tend to increase costs.

My last issue is the question of international equity. There are really two distinct issues here: equity between rich and poor countries, and equity among industrialized countries.

Poor countries are quite right in noting that they have much lower per capita emissions now than developed countries. Poor countries also have very pressing development needs. For either or both of those reasons, these countries will be disinclined to participate in any emission reduction scheme that they have to pay for. Without their participation, OECD-only emission reduction efforts are more symbolic than real in terms of affecting atmospheric concentration of greenhouse gases given the statistics I cited earlier.

The question then arises, would the richer countries be willing to pay for emission reduction costs in developing countries as well as at home? Assuming that they would, could developing countries make better use of the same resources for some other purpose?

Finally, even if poor countries are not asked to make any emission reductions, we have to recognize the highly interdependent nature of the world economy. There is a clear relationship between economic growth in the developing countries and economic growth in the OECD. Emission reduction policies, even if restricted to the industrialized countries, could have an adverse impact on poorer countries.

Let me turn quickly to the question of equity among industrialized countries. The tendency to view all industrialized countries as being in the same boat is probably misleading. OECD countries differ dramatically in may ways. For instance, the United States expects about thirteen percent growth in population over the next twenty years. The European Community expects two-percent population growth over the same period. In terms of resource endowments, the U.S. has abundant low-cost coal and limited natural gas. Europe has high-cost subsidized coal and access to lots of natural gas. These differences can have a big impact on the evaluation of emission reduction objectives and their costs. Therefore, such objectives must be carefully analyzed on a country or area-specific basis.

IMPLICATIONS FOR DECISION MAKING: THE ELECTRIC UTILITIES' PERSPECTIVE

William L. Fang

Assistant General Counsel
for Industry Affairs
Edison Electric Institute
701 Pennsylvania Avenue, N.W.
Washington, DC 20004

I will present the electric utilities' perspective on this issue in three parts. First, I will address an economic study that is being conducted for our industry. Second, I will discuss the interrelationship of energy, electricity and the environment, and the global climate change issue, and finally I will conclude by addressing the global context of the issue.

We have commissioned ICF Resources, Inc., a well-known consultant on energy/environmental issues, to conduct a major assessment of the economic impacts of greenhouse gas policies for our industry and our customers in terms of costs, impacts and opportunities. It is a draft study currently. It is undergoing review and revision, so I will just be presenting a preliminary overview. The study is a micro-economic examination of our industry focusing on CO_2 emissions primarily. The model is a very complex, sophisticated, national planning model which integrates supply-side and demand-side options.

Basically there are high-impact and low-impact reference scenarios or base cases. All of the policy scenarios are being run against these two base cases. Some parameters are price elasticity, electricity demand, and new nuclear additions. Both of the reference scenarios are realistic. The high-impact scenario looks somewhat like today. No nuclear is allowed, but there is significant penetration of renewables and significant reliance on natural gas. And the low impact scenario also looks somewhat like today, but you have significant nuclear capacity coming on-board. Renewables make even more significant contributions and natural gas is heavily relied upon.

There are four policy cases: (1) stabilization by the year 2005, (2) stabilization by 2005, then a 20-percent reduction by 2015, and (3) a 20-percent reduction by 2005, held constant thereafter.

Global Climate Change, Edited by J.C. White
Plenum Press, New York, 1992

We also look at a carbon tax scenario, the fourth policy case. This is a phase-in tax that would start at $11 per metric ton in 1996, then increase $11 per year and reach $110 per metric ton by the year 2005.

There are thirteen regional submodels that comprise the national model. Some of the key assumptions the model will try to achieve are least-cost, cost-effective, and utility-wide solutions across the country.

Some other important assumptions are: (1) no trading will be assumed internationally, (2) no trading will be assumed between utilities and other sectors of the U.S. economy, and (3) no trading will be assumed between CO_2 and other greenhouse gases. There is implicitly trading assumed among the electric utility industry itself on CO_2.

Carbon emissions are going to grow substantially under both reference scenarios. The magnitude of this might change somewhat, but the slope of the curve is unmistakable, rising to the years 2005, 2015 and beyond.

In the base case generating mix, we have demand-side management (DSM), imports and purchases from Mexico and Canada and elsewhere, renewables, nuclear, gas, coal and oil. Coal consumption increases substantially in the base case. Gas usage will rise sharply at first and then go downward. The study is probably going to indicate that increases in the price of gas will take a tremendous toll and demand for gas will increase in all sectors of the economy.

Some tremendous reductions in carbon emissions will be necessary to get to stabilization or a 20-percent reduction.

Several strategies will have to take place in order for reductions to occur. There are price effects—that is, reduction in electricity demand because of the higher prices—DSM, renewables, nuclear, fuel switching to natural gas, and energy efficiency improvements. Coal consumption is going to go down.

Finally, we asked ICF to look at greenhouse gas removal costs in other sectors. What are some of the other ways to reduce carbon emissions other than requiring our industry to do this on our own? What else can be done? Some possibilities are collecting CFCs from refrigerators, reforestation, collecting coal bed methane, and capturing the methane from new and existing landfills.

In the second part of my presentation, let us look at the interrelationship between energy, electricity and the environment. Let us look at the relationship between electricity and economic growth in the U.S. between the years 1973 and 1990: basically electricity and GNP are in tandem. There is a much less distinct relationship between electricity and energy demand. There is a continued reliance on electricity which is critical to a healthy economy, but energy use is not as important to a growing GNP.

Over the same time period—1973 to 1989—electricity demand has increased in the U.S., while carbon emissions per unit of GNP have decreased. In other words, overall carbon emissions may be increasing in the U.S. due to economic growth, but carbon emissions per unit of GNP have actually decreased over the last sixteen years. The economy has become increasingly reliant on electricity, but at the same time has become less carbon-intensive.

You might ask how this is possible. One of the explanations is that there is an increasingly widespread use of electricity in new end-use applications and also an increasing substitution of electricity in existing end-use applications. In various sectors of industry, commerce and transportation, electricity has become the end-use fuel of choice because of its environmental cleanliness, its efficiency and its precision. The microwave oven, the fax machine and, of course, the coming electric cars are some of the end-use electric technologies that are displacing fossil fuel energy forms. They can, of course, benefit the environment.

I would like to note that the electric utility industry has also become less CO_2-intensive over the last nineteen years. In other words, power plants are emitting less CO_2 per unit of electricity sales than they used to.

Turning to the international scene, the global climate change issue must be placed in a global context before this country commits itself to massive or costly mitigation options. Greenhouse gas emissions are projected to rise from about 11 billion metric tons of carbon equivalent to nearly 28 billion metric tons by the year 2050. To stabilize greenhouse emissions would require a tremendous reduction in the projected levels. Or, looking at it another way, stabilization of greenhouse gas emissions would imply no growth in many countries and perhaps negative growth in other countries.

You have an even tougher scenario if the goal is to stabilize atmospheric concentrations of greenhouse gases. That kind of policy would require massive reductions, severe economic dislocations and probably negative growth in most countries.

World population is expected to grow to perhaps ten or even twelve billion people by the middle of the next century. While the world is currently about 40-percent urbanized, it could then be as much as 70-percent urbanized.

The outlook for world energy utilization using coal, renewables, natural gas and uranium all go up. The only one that is going down is oil.

It does not take a rocket scientist to determine that the countries with the largest coal reserves, which are China, the former U.S.S.R. and the U.S., will be relying on fossil fuels as a major source of energy in the coming years.

The economic losses which occur under a carbon limit produces winners and losers. The biggest loser is China. The former Soviet Union, Eastern Europe, and the U.S. are obviously going to sustain heavy economic losses. The OECD is clearly one of the winners.

For the U.S., there are two points to make. The U.S. would be affected more quickly than our international trade competitors and we would be affected about twice as much as the OECD or our international trade competitors.

In conclusion, let me say that, before the U.S. embarks on any program that embraces carbon emissions limitations, policymakers must comprehensively examine both the international implications of our actions and the impacts on the U.S. economy. In addition, policymakers should also carefully consider the ramifications of such proposals in limiting electricity demand, which could, in turn, negatively affect economic growth and the environment.

IMPLICATIONS FOR DECISION MAKING: AUTO INDUSTRY PERSPECTIVES

Samuel A. Leonard

Director
Auto Emissions Control
General Motors Technical Center
30400 Mound Road
Warren, MI 48090-9015

The first reaction of the auto industry to the global warming issue is that we wish it would go away. We have enough problems without it, and if you add anything to recession and foreign competition, you will make things much more difficult. But obviously the threat of global warming isn't going to go away so the second reaction is: is this global climate thing real or imaginary? I must admit that my own personal opinions are much closer to the imaginary side than those of our keynoter, Dr. Oglesby.

But on further reflection, it really doesn't matter because as far as the auto industry is concerned, there are a lot of other forces pushing in the same direction. It doesn't make any difference whether it is global climate change, whether it is balance of trade, whether it is energy security—all of the pressures are to make the vehicles more fuel-efficient or, in some other way, to reduce the usage of fuel by motor vehicles.

Then we get to the question of, if this is going to happen, if fuel usage by the transportation sector is going to be reduced, how would we prefer to see it done? There are two classic modes of doing it: (1) command and control by regulations and (2) market-driven mechanisms. The primary driving force for the command and control example over the last fifteen years has been the corporate average fuel economy or, as we call it, the CAFE standards.

We have proposals in Congress today which would require significant increases in the required fuel efficiency of automobiles (Figure 1). There is a question in our minds whether we can meet the new standards and still sell the product to the consumer. Aside from that, do they achieve the objectives of command and control which they are asserted to achieve?

Go back to 1975 and consider the basic claims which were made for the CAFE law. The proponents of the law claimed that the CAFE law would reduce our dependency on foreign oil and reduce our oil consumption by 2.5 million barrels per

Global Climate Change, Edited by J.C. White
Plenum Press, New York, 1992

day. The opponents of the CAFE law, the auto industry, said it would force everybody into smaller cars such as Pintos and Mavericks. Which of these claims were correct? The dependency on foreign oil went from 35 percent in 1974 to 42 percent in 1990. The usage in barrels of oil per day went from 6.5 million barrels of oil per day in 1974 to 7.2 million barrels per day in 1990. Neither one of those claims was proven true. The average 1990 car has the exterior dimensions of a 1974 Pinto wagon and the interior dimensions of a 1974 Ford Maverick.

	Bryan (1988 Base)	*Johnston* (1990 Base)
Pass. Cars	**20% by 1996**	**7.8% by 1996**
	40% by 2001	**21.4% by 2001**
		32.1% by 2006
Trucks	**20% by 1996**	**6.3% by 1996**
	40% by 2001	**16.4% by 2001**
		28.5% by 2006

Figure 1. New CAFE Proposals.

These same claims are being made for the present CAFE round, and again we think that, in order to meet the CAFE standards, the vehicles are going to have to get smaller. The laws of physics do not allow you to do anything else.

CAFE has additional side effects to go with it. By forcing unrecoverable costs on the new vehicle, you reduce new vehicle demand. That slows the turnover of the fleet, keeps the older vehicle on the road longer, or forces the people into categories of vehicles that are not automobiles, such as trucks and vans, a trend we have seen in the last fifteen years. It causes no reduction in the amount of vehicle miles traveled. People tend to drive just as much, especially with the price of fuel the way it is, and by slowing the turnover of the vehicle fleet, due to the higher new vehicle prices, you also prevent reductions in emissions of other pollutants—hydrocarbons, carbon monoxide and oxides of nitrogen.

What about the market-based mechanism? The primary one that I am going to look at is energy taxes. You can call them carbon taxes, energy taxes, or gas taxes. One form or another, they are energy taxes. Each one of them has its own positive side and negative side, but let us just talk about energy taxes in general. Energy taxes are a simple application of a supply-and-demand curve. If you increase the cost, the demand goes down. I learned that in Economics 101 and I am an engineer, not an economist.

Let us look at the cost of driving versus the average miles per year a person drives in dollars per 10,000 miles (Figure 2). There is an inverse relationship between how much it costs to fuel a vehicle and how many miles it is driven every year—almost mirror images on the graph. This particular chart includes the effect of the improved mileage of the vehicle, that is the cost per mile, not the cost of fuel alone.

If you look at the cost of fuel alone in dollars per gallon (Figure 3), you see also a very similar relationship in dollars per gallon and vehicle miles traveled. Over the time span of 1974 to 1990, we have seen the miles traveled per year per vehicle go from a high of around 10,100 miles per year in 1988 when gasoline, combined with fuel economy, put driving at its all-time low cost per mile for energy, to as low as 9,200 in 1980 right after the 1979 oil embargo when the price of gasoline was quite high.

Energy tax also has side effects, but generally for the auto industry these side effects are positive. It increases new vehicle demand, because people want to take their twelve mile-per-gallon (MPG) clunkers and get rid of them and buy a new car that gets 28 or 29 or 30 MPG. It reduces the vehicle miles traveled (VMT) which has an added effect of reducing the emissions, and both the reduced VMT and the turnover of the fleet accelerate reductions of the other emissions that this country is interested in controlling, such as carbon monoxide and oxides of nitrogen.

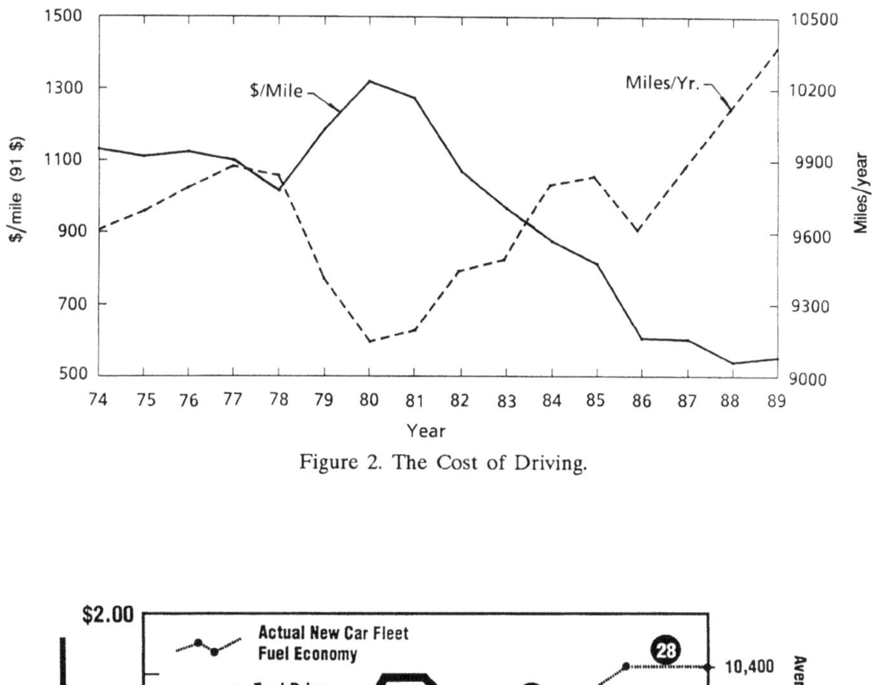

Figure 2. The Cost of Driving.

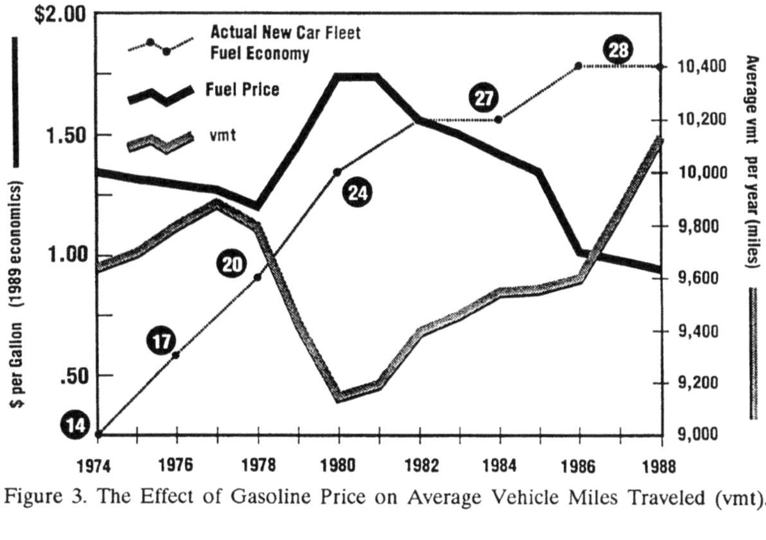

Figure 3. The Effect of Gasoline Price on Average Vehicle Miles Traveled (vmt).

In conclusion, the auto industry thinks that if control has to happen, the market-driven mechanisms are much superior to the command and control mechanisms and would prefer to see those mechanisms as opposed to additional CAFE requirements.

IMPLICATIONS FOR DECISION MAKING:

INDUSTRIAL SECTOR PERSPECTIVES

Frederick E. Mangelsdorf

Manager
International Environmental Programs
Texaco, Inc.
P.O. Box 509
Beacon, NY 12508

The industrial sector in the United States is sometimes painted as being inconsiderate, insensitive, inept or incorrect with regard to environmental issues. By and large, none of these adjectives is correct. This is particularly true with regard to the subject of global climate change and the implications for decision making. This paper will review some of the major points that reflect an industrial viewpoint of the climate change issues.

The topic of global climate change is very important and the decision making process too important to be left only to the politicians, or to industry, or to the scientific community, or to the environmental community, or to the press. Indeed, the population as a whole must be involved and must participate in the decision making. Those in society who are more versed in this subject than others have a responsibility to bring others into the dialog and to ensure that all sides of the story are presented.

The press has a special responsibility to present the facts and to resist the temptation to present new claims without verifying their credibility. There is a responsibility also to present new facts in the context of what is already known. And with this in mind, the press should take full advantage of the advisory services available to them, one of the best being the Media Resource Service of the Scientists' Institute for Public Information. There is scientific advice available, and it is free.

As a people, as a nation, as the world community of nations, we are faced with significant decisions about our lifestyles, about our relationships with one another, and about our responsibilities to one another. Indeed, we are facing crossroads which require the very best decisions we can make. We must search for the truth and for the best answers possible.

Global Climate Change, Edited by J.C. White
Plenum Press, New York, 1992

Unfortunately, there are very few easy answers. For instance, the science is uncertain. Our scientific knowledge is growing, but things that were regarded as gospel a few years ago, or only a few months ago, have been refuted by new knowledge. And it is only logical to assume that some of the things we regard as gospel today will at some time be refuted. This is indeed the nature of science.

Scientists look for explanations of physical facts—for cause-and-effect relationships. Unfortunately, there is not yet a good explanation of cause-and-effect relationships with regard to observed temperature changes in the world, and thus to the whole climate change issue.

The global climate models are uncertain. There are honest differences of opinion between scientists of great repute as to the validity of the construction of the models, and as to the interpretation of the results. Many of us are left in doubt as to what the models are really telling us.

The consequences of warming are uncertain. It probably will not become uniformly warmer over the entire globe. Rather, some areas may actually cool, others may see little change, and some areas may see an increase in temperature. But where? And how much? Again, no easy answers.

The economics are uncertain. All of us have heard about or read about the wide range of answers being produced by the different economic models. There is indeed uncertainty and healthy dialog in the economics community as to the cost of measures to reduce carbon emissions and of the long-term impact on the economy of adopting such measures.

Given all the uncertainty about the science, the models, and the economics, one would hope that there would be certainty about the technologies to mitigate or adapt to climate change. But such a hope would be in vain. For example, one of the mitigating measures that is discussed in the NAS Mitigations Panel Report is that of de-tuning jet aircraft engines that are flying at high altitudes so that the fuel is not all burned completely, but rather the engines would emit some soot from their tailpipes. The theory is that this soot in the upper levels of the atmosphere would reflect the sun's rays, thus providing an offset to the expected increase in greenhouse effect from the greenhouse gases. And this could be done at a cost of only pennies per ton of CO_2 effect that was abated. When one looks at some of the economics papers suggesting that the cost of abating CO_2 emissions will be measured in hundreds of dollars per ton, a strategy that might cost only pennies per ton looks pretty good. But think of the uncertainties. Do we really want to try to counteract one of man's potentially harmful emissions with another emission? From a cost-effectiveness standpoint it certainly looks attractive, but one has to have doubts that this would indeed be the right thing to do.

In spite of all the uncertainties, however, there are some things that can and should be done. One of them is to implement the Administration's proposed "no regrets" policy, to reduce greenhouse gas emissions when it makes sense to do so for reasons other than global warming. And conservation is one of them.

Let me assert that a lot of energy in this country is wasted. Who among us has not lived at one time in an apartment, or worked in an office, where there was no thermostat and in which it got so hot that the windows had to be opened in the winter in order to cool off? Who among us does not turn off the lights in our home or office when we leave the room, even for a few minutes? Who among us does not set the thermostat down at night, or during the day when we will be out of the house for several hours or more? I submit that these things should be done. We should have a conservation mind-set, and we should practice it and preach it.

There has been much discussion of late about a U.S. National Energy Strategy. But what, one might ask, has been the national energy strategy for the several hundred years our nation has existed? We really haven't had a formal one, but the informal, unstated one seems to have been that cheap energy is good for the Nation. When the first European settlers arrived, there was firewood to be had for the cutting. There were creeks to be dammed for the installation of water wheels, and coal to be had for the digging. There were the winds of the prairies to turn windmills. There were immense rivers to be dammed to provide cheap electrical power. And so on. The nation thrived on cheap energy. We grew to the point that we have today one of the highest per capita GNPs in the world.

I'm not about to knock success. But I do think that the amount of energy that we use can be reduced while maintaining a high standard of living. You may have noted that my company is in the energy business. Yes, we sell energy, and yet I'm saying to all of you, "Let's reduce the amount of energy that we use. Let's not waste it." We believe in conservation. We practice it. And it has been a recurring theme of our advertising. We believe that energy is too precious a resource to waste.

As we move forward and address the climate change issue, industry can and will participate in the debate and in the quest for solutions. Industry is concerned about environmental problems and about sustainable development for all the peoples of the world.

Earlier this year at the World Industry Conference on Environmental Management, a Business Charter for Sustainable Development was adopted. The Business Charter identifies some sixteen action steps to be taken by companies to make the world a better place as it continues to grow and develop. To date this Charter has been endorsed by more than 360 companies around the world.

There are numerous industry associations and action groups working on the issue of global climate change. The groups with which I am familiar have objectives to elucidate the facts regarding the science and the economics, and to participate in the public dialog about the problems and potential solutions.

Obviously, global climate change isn't just a U.S. problem, or a Northern Hemisphere problem, or a western problem, or an OECD problem, or a problem only of the developing countries. It is a world problem. Yet, the world political structure seems ill-equipped to deal with the issue on a global basis. There is much to do to build the trust, the communication links, and the knowledge of one another's cultures, needs and desires, that will allow us to arrive at mutually beneficial solutions.

The chart in Figure 1 illustrates the global scope of the problem. It is a plot of projected global CO_2 emissions from anthropogenic sources in billions of tons of carbon per year, for the years 1987, 2020, 2050, and 2100; it was prepared earlier this year by the Global Industrial and Social Progress Research Institute in Japan.

The upper part of each bar is for the developing countries; the middle part is for the centrally planned economies; and the lower part is for the developed countries. The assumptions for the future projections are these:

- That the developed countries would stabilize total CO_2 emissions at 1987 levels.
- That the centrally planned economies would have an increase in per capita CO_2 emissions of fifty percent by the year 2050.
- That the developing countries would have a doubling of per capita CO_2 emissions by the year 2050.

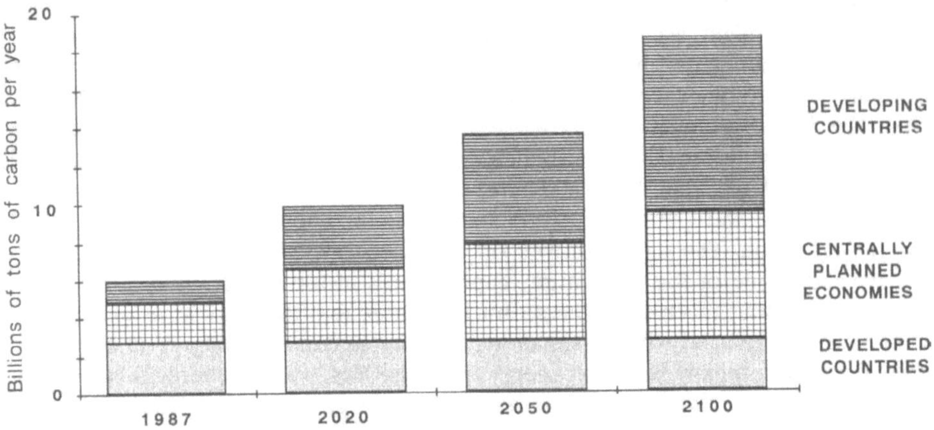

Figure 1. Global CO_2 Emissions.

If these assumptions are at all reasonable, it becomes very clear that the developed countries can't solve the problem alone even if Draconian steps are taken to stabilize CO_2 emissions at current levels. There will have to be a world solution that involves all nations. The graph also suggests that, on a world basis, it looks highly likely that there will be a significant increase in CO_2 emissions in the future. And this leads to the conclusion that we should begin to think very seriously about how we are going to adapt to a world with substantially higher CO_2 levels.

There are very real differences in perspective with regard to climate change issues. Developing countries sometimes view the developed countries as being profligate with energy use. Some European countries believe that the developed countries, and particularly the U.S., should impose carbon taxes immediately. The U.S. points out that on the basis of energy use per unit of GNP it is not profligate—other countries do not do as well as does the U.S.

Some countries are viewed as being profligate in population increase. And there is no question that the more people there are, all of whom want to use more energy, the greater will be the CO_2 emissions from fossil fuel combustion.

There are other differences in perspective. Religion, culture, language, trade relations, historical alliances and so on ensure that there will be many different perspectives on this issue. This doesn't mean that we can't agree on an assessment of the future, or on a program of what to do about such an assessment. But we shall all have to develop a keen sense of recognizing and respecting other people's perspectives if there is to be a true dialog on the issue of climate change.

I'm very pleased to be a representative from industry. There is much that industry can do to help:

- With the dialog.
- With knowledge and best guesses about technology
- With the implementation of action plans to carry-out government initiatives.
- With best estimates as to the cost of undertaking different action plans.
- With information about trade-offs and how these may affect the economics of different action plans.

To the extent that it is feasible to do so, industry can benefit from having clear guidance from society as to society's expectations for industry. And it is entirely appropriate that society can expect full and complete cooperation from industry in reaching the objectives that are set by society.

There is no free lunch however, and it must be remembered that a company is not different from organizations like churches and universities—if its revenues are not greater than its costs, it shall soon cease to exist. To say it another way, if there are costs to be borne by industry in reducing CO_2 emissions, those costs will have to be borne by the customers through higher prices for the goods and services provided.

In brief, industry perspectives can be summarized in a few short statements.

- The topic is very important.
- There are very few easy answers.
- There are some things that can be done.
- Industry can and will participate.
- This is not just a U.S. problem.
- There are very real differences in perspective.
- Industry can help.

With these perspectives in mind, industry is ready and willing to assist as the world deals with climate change issues.

DISCUSSION

IMPLICATIONS FOR DECISION MAKING: SECTORAL PERSPECTIVES

QUESTION: Can you give us any guidance as to how one could reduce the impact on the electric industry?

FANG: I think basically our industry favors a research and action agenda. Obviously, there are a lot of uncertainties—scientific, technical and economic—and so, certainly, the $1 billion that the federal government is spending on an annual basis for scientific and economic research is well justified. We are spending money in the private sector through the Electric Power Research Institute and conducting other studies of our own.

For the action part of it, we believe that the "no-regret" strategy or what some people have called the "no-regrets or no-regrets-plus" approach makes sense. For our industry that means focusing on no-cost and low-cost options at this point, considering the demand-side and supply-side options together.

Dr. Gruenspecht mentioned integrated research planning. Some of our companies are way ahead on the demand-side options. Certainly we favor energy efficiency, electricity conservation, demand-side management (DSM) programs. On the supply side, we favor anything that can be done to improve power plant efficiency and heat rates, to improve the capacity factors for nuclear plants, and to improve distribution transformers. We are trying to push our industry into more efficiency.

QUESTION: Would oil consumption in the USA be as low as it is without CAFE standards?

LEONARD: In 1982 we were predicting that the CAFE of our fleet would be above 30 miles per gallon, based on a projection of the oil prices being above $2.50 per gallon. I think the improvement you are seeing in the fuel economy has been driven largely by the price of oil to this date, rather than by the CAFE standards themselves. In the late 1970s and early 1980s we were well ahead of the required CAFE fuel economy levels, so it was market-driven by fuel prices.

QUESTION: China and India have reiterated that they intend to fully exploit their indigenous coal resources to advance economic development. If they do, this will more than negate other efforts on CO_2 control. Who, if anyone, will regulate them?

GRUENSPECHT: I think that the question reiterates the point that I tried to make, that in looking at emissions, you have got to look at global emissions. I don't think anyone is in a position to regulate them. I don't know on what basis regulation imposed from outside would be justified. I think the UN system frowns pretty heavily upon infringements on sovereignty.

Again, this raises the issue of our objective here. Given climate sensitivities, and given the link between national emissions and global emissions, and given the link between global emissions and atmospheric concentrations, and then given the link between atmospheric concentrations and climate, the benefits of emission reduction policies are uncertain. We say that maybe those benefits are more important in the developing countries, and but the developing countries say that, relative to their other needs, these things may not be their highest priority.

A question like this leads you to question the a priori notion that this is an appropriate first-tier objective for mankind. One could think of a regime in which trade restrictions are applied as a means of coercing emissions reductions in other countries. But one has to ask if we are really benefiting mankind as a whole by doing that if, indeed, these countries have other development priorities and trade is important to their development priorities.

This raises the issue of looking at where are we expecting to get benefits and what policies we should pursue. An ultimate solution may be technology cooperation. If technologies are inherently economically more attractive, they will be adopted by India, China and everybody else.

MANGELSDORF: I would like to expand on that. Clearly, we, mankind, are getting to the point where we probably technically can control earth's temperature. We think that by putting up more CO_2 we can increase the temperature, by putting soot up in the upper atmosphere we could decrease earth's temperature. We don't have a political system that allows us to make important decisions, such as, what is the right temperature. Who is going to decide? Secondly, who gets to put their hand on the thermostat and change it and, thirdly, what do we do if somebody unilaterally changes it without agreement of world society? The technology is beginning to be there, just as technology has gotten in front of our ability as people to sort out some of these issues. How do we build a world political structure that allows us to debate this kind of an issue and come to conclusions?

QUESTION: There is some evidence that nonhumans may be affected quicker by global warming than humans. The question is, to what extent do DOE, Ford, Texaco and the electric industry consider the impact on the other 30 million species with which we share the planet? Does that play any role in your considerations?

MANGELSDORF: The short answer for us is yes.

LEONARD: I have no idea how Ford feels, but it is one of the considerations that everybody has to look at, I think.

GRUENSPECHT: It certainly plays a role for us, but since only people vote, the issue comes down to how people view those impacts. But I think people take those impacts very seriously and those are critical areas of benefit.

QUESTION: The auto industry wishes to let the market dictate what happens, supply and demand, and yet recently the big three auto makers turned to Washington when Honda announced it has the capability to make cars that are more fuel-efficient. In light of your comments today, how do you justify such actions?

LEONARD: Honda's lean-burn engine is not a technological break-through. It won't meet future emission standards. But if you are talking about going to Washington on the "dumping" question, that is a totally different issue. That is unfair competition. It is selling below cost in this country, which is against the international trade rules.

QUESTION: How can you justify increased use of electricity when much of this projected use would be in heating, cooling and other inappropriate uses of such a high quality energy?

FANG: I think my point dealt with increased electrification and use applications. Some of those could be in heating and cooling but, with modern technical advances such as highly advanced heat pumps, advanced forms of water heaters and that kind of thing, efficiency is so improved that I am not sure I agree with the premise of that question.

I am not saying that all end-use applications of electricity are beneficial for the environment or will reduce CO_2 emissions. Clearly there are a number in the transportation, industrial and commercial sectors—such as electric vehicles, fax machines, microwave ovens and so forth—that should achieve much higher penetration in the future.

QUESTION: Could you explain why the U.S. would be adversely affected relative to OECD countries if CO_2 emissions were reduced? You showed a slide which indicated that the U.S. would be hurt more. Could you amplify why that is the case?

FANG: I will try to address that, but Dr. Gruenspecht might want to address that too. I am not an international trade expert. Just the constraints of carbon limits are going to hit our country a lot harder than the OECD countries, whether that is under stabilization or a 20-percent reduction.

GRUENSPECHT: I tried to address that a little bit. Part of it has something to do with conditions. Look at expected population growth. If you look at the resource base and keep in mind that, even though there is trade in energy (remembering that coal and natural gas have very high transportation costs relative to their value), where they are located really matters a lot. Fuel switching options look better in Europe than they look in the United States.

There is no reason to expect the industrialized countries all to be in one boat. That is an important over-simplification. Furthermore, even within Europe, there is a lot of controversy between the northern and the southern European countries. They don't view themselves as being all in one boat and are jockeying for position among themselves.

Implications for Decision Making

Societal Considerations

THE IMPORTANCE OF POPULATION GROWTH IN FUTURE COMMERCIAL ENERGY CONSUMPTION[1]

Gretchen Kolsrud

Senior Associate
Office of Technology Assessment
United States Congress
Washington, DC 20510

Barbara Boyle Torrey

Chief
Center for International Research
U.S. Bureau of the Census
Washington, DC 20233

INTRODUCTION

Today, global commercial energy consumption contributes to pollution, environmental deterioration, and global greenhouse gas emissions. Increases in global commercial energy consumption are driven by population growth and economic development, which tends to increase energy use per capita. Therefore, the inevitable increase in population in the near term and the economic development that must necessarily occur in many countries poses serious implications for the environment [OTA, 1991].

Many investigators have considered the relative contribution of population growth to environmental deterioration [Keyfitz, 1990]. Some find population growth to be highly detrimental [Ehrlich and Ehrlich, 1990; Sadik, 1990], while others find population growth to be detrimental in the short-run but of little consequence in the long-run [Shaw, 1989]. This paper addresses one aspect of this issue. It estimates the contribution of population growth to commercial energy consumption, which is considered a major, although not the only, cause of increases in air pollution and greenhouse gases today.

Global Climate Change, Edited by J.C. White
Plenum Press, New York, 1992

This paper first summarizes some of the recent estimates of future energy use developed by well-known models. It then develops several alternative scenarios that use different assumptions about population growth and energy use per capita for 122 countries for the years 2020 and 2050. It calculates the relative contribution of population growth to the change in total commercial energy use and demonstrates the sensitivity of the results to different assumptions.

Individual country data are separately summed to totals for more-developed countries (MDCs) and less-developed countries (LDCs). Under a business as usual scenario for both MDCs and LDCs, population growth is important, but not the most important factor, in future increases in global energy consumption. Analysis of other scenarios shows that while slower population growth always contributes to a slowing of future global energy consumption, such changes are not as effective as reductions in per capita commercial energy use.

Calculations on a global basis are made in two ways: from global aggregates and by summing individual country data. Comparison of the results shows that the first method is misleading because of the heterogeneity of population growth rates and energy consumption rates of individual countries.

The tentative conclusions reached in this paper are only small pieces of a much larger puzzle. More work needs to be done to better understand the dynamics of these relationships before the analysis is extended to the broader questions of population growth and environmental change.

A BRIEF REVIEW OF FOUR MODELS

A number of models in the last two decades have used population growth and other parameters in their efforts to estimate future increases in global energy consumption. Their major purpose was to estimate increases in global energy, not to determine the contributions of population growth to such increases. Nevertheless, the results of their models demonstrate the relative importance of population changes in the energy equation.

The results of using four different models to generate six scenarios of future energy consumption are presented in figure 1. A discussion of the models and their strengths and weaknesses can be found in *Scenarios of Socioeconomic Development for Studies of Global Environmental Change: A Critical Review* [IIASA, 1989]. What is of interest for this paper is the relative importance of future population increases. All six scenarios use the same population projections; the differences among them arise from varying assumptions regarding the rate of economic development, the relationships between economic development, and per capita energy use and the resulting global total energy consumption [IIASA, 1989, p. 97].

The projected increase in global energy consumption from 1975 to 2000 varies from 3.8 terawatts (TW) in the Nordhaus and Yohe model to 8.7 TW in the IIASA high projection. Between 2000 and 2025, the difference in the change increases from 3.1 TW in the WEC low model to 15.2 TW in the IIASA high projection.

As can be seen in figure 1, in all scenarios, growth in population is a larger component of change in total commercial energy consumption than growth in per capita energy consumption from 1975 to 2000. But between 2000 and 2025, the increase in energy per capita is a similar or larger factor than population growth in explaining the growth in global commercial energy consumed, except in the WEC low model, which assumed no change in energy per capita.

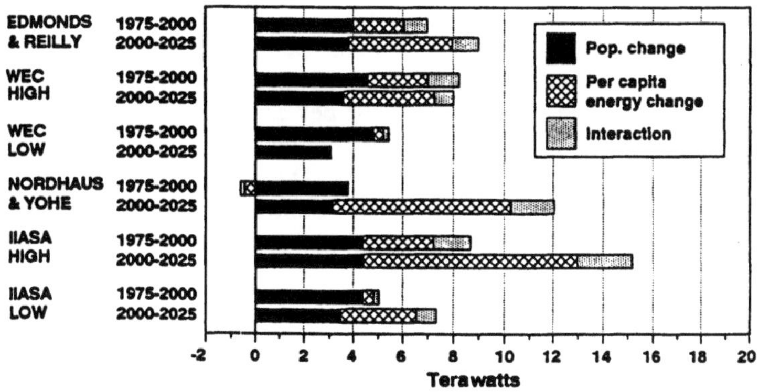

Figure 1. Components of Change in Commercial Energy Consumption Projections, 1975 to 2000 and 2000 to 2025.

These estimates were done in the early 1980s using regional aggregates for population, economic, and energy inputs. They assume slower population growth than has occurred. (The population projections used assumed that world population was 5.9 billion in 2000 and 7.4 billion in 2025. In fact, world population now is estimated to reach 5.8 billion by 1995 and be 8.7 billion in 2025.) And they also assume that energy per capita continues to increase in the developed countries. To explore the implicit conclusion of these models about the relative importance of population growth, this paper uses current population figures and more recent energy per capita consumption data for 122 individual countries rather than regional aggregations. These are the countries for which there are recent, reliable population, economic, and energy data. The time horizon is 60 years.

ASSUMPTIONS

Population Growth. World average population growth rates peaked in the 1960s at about 2.0 percent a year. Today, world population is growing at 1.8 percent a year. Medium population projections expect population growth to decrease slowly in the future until a population of 11 billion is reached in about 2060. This assumes that total fertility rates in Africa begin to decline in the 1990s or continue if they have begun to decline, that rates in the developing countries continue to decline, and that rates in the developed countries stay below or near replacement level.

The population projections used in this paper were produced by the Center for International Research at the U.S. Bureau of the Census on an annual basis [U.S. Bureau of the Census, 1989]. They incorporate the latest available data on fertility, mortality, and migration. Projections of these components were made separately and integrated. The resulting estimates are consistent with those made by the United Nations and the World Bank, although individual country estimates may differ from the projections made by those agencies. They are less optimistic than the ones used by the models discussed above.

Energy Consumption Per Capita. From 1965 to 1988 the world's real GNP per capita grew at an average annual rate of 1.5 percent [World Bank, 1990]. The middle-income economies and high-income economies combined grew at 2.3 percent on average, which means that their real GNP per capita doubled in 30 years. India and China grew more rapidly, at an average annual rate of 4.0 percent per capita, whereas the poorest countries grew less rapidly. The assumption made in this paper is that GNP per capita will double in each country from 1988 to 2020 and will double again from 2020 to 2050. Obviously, some countries will do less well; others will do better. GNP growth is relevant to this paper because energy consumption per capita is generally correlated with it.

Today, per capita energy consumption is about 1 gigajoule per person in many African countries compared to 280 in the United States [WRI, 1990]--a range of nearly 300 to 1. Comparisons of changes in world energy consumption per capita and GNP show that the two are positively correlated but may not change at the same rate [IIED and WRI, 1987]. Economies become more efficient as they develop because of advances in technology and changes in behavior. And as countries move from using non-commercial forms of energy to commercial forms, the relative price of energy increases. Moreover, economies go through structural changes from heavy manufacturing to lighter manufacturing and services, and these changes affect per capita energy consumption. In the following scenarios, GNP is assumed to double in each country in each 30-year period, but different assumptions are made about the relationship of energy consumption per capita to GNP during the same periods.

SCENARIOS USED IN THIS STUDY

Five scenarios are the basis for the analysis reported here. Briefly described, they are:

1. Unconstrained Scenario. In MDCs, energy consumption per capita increases by 50 to 80 percent in each 30-year period and in LDCs increases by 80 to 100 percent during the same time. Medium population projections are used.

2. MDC Energy Constraints Scenario. MDC energy consumption per capita is constrained to 1988 levels until 2020 and is cut by 10 to 30 percent between 2020 and 2050. LDC energy consumption per capita increases 80 to 100 percent in each period. Medium population projections are used.

3. MDC Energy Constraints Scenario with Lower LDC Population Growth. Scenario 2 with the addition of a 20 percent reduction in baseline LDC population growth in each 30-year period.

4. Tougher MDC Energy Constraints Scenario. MDC energy consumption per capita cut 10 to 30 percent between 1988 and 2020 and again between 2020 and 2050. Medium population projections are used.

5. Tougher MDC Energy Constraints Scenario with Lower LDC Population Growth. Scenario 4 with the addition of a 20 percent reduction in LDC population growth in each 30-year period.

Details for each of the five scenarios are provided below. Readers may wish to skip to the Results and return to the details later.

1. Unconstrained Scenario. In the post-war period prior to 1972 and from 1986 to 1988, U.S. energy use increased at the same rate as GDP [OTA, 1990]. This is unlikely to continue indefinitely. Therefore, this scenario begins with the more conservative assumption that in the next 30 years, GDP doubles but energy use per capita rises by 50 percent in countries with high per capita energy use today. Further assumptions are detailed below.

> **More Developed Countries (MDCs) and Organization of Petroleum Exporting Countries (OPEC).** Energy consumption per capita (EC/C) increases by 50 percent in each 30-year period for countries consuming 220 or more gigajoules per capita (double the present-day level of Japan). When EC/C is less than 220 gigajoules, EC/C increases by 80 percent.
>
> **Less Developed Countries (LDCs).** EC/C doubles in 2020 and again in 2050 if 1988 EC/C is equal to or less than 20 gigajoules and total fertility rate (TFR) is less than 5. Under all other conditions, EC/C increases by 80 percent.

2. MDC Energy Constraints Scenario. Under this scenario, per capita energy consumption in the developed world (38 countries) is constrained to present levels and then lowered, with the reductions being greatest for those countries with the highest levels of per capita energy use. The "developed world" is defined as those countries with per capita energy consumption at or above 110 gigajoules per person, the present level in Japan. Countries now consuming at lower levels are allowed much more room for growth in energy consumption, only slightly less than in the business as usual scenario.

There is evidence that increases in energy conservation and in energy efficiency at levels needed to support this scenario are technologically feasible without compromising economic growth. Between 1972 and 1985, U.S. GDP grew by 39 percent while energy use remained virtually unchanged [OTA, 1990]. One-third of the U.S. decline in energy intensity during this period was due to a change in the industrial composition of the economy. This indicates that there is considerable room for reducing energy use without compromising economic growth in the developed world.

> **Countries at or above 110 EC/C** (present-day level in Japan; those at or above this level generally are MDCs and OPEC). EC/C held at current levels to the year 2020. For year 2050 estimates, assume countries above 200 EC/C cut the per capita energy use 30 percent; between 150 and 200 EC/C, cut 20 percent; and between 100 and 150 EC/C, cut 10 percent. Those below rise to 100 EC/C by the year 2050.
>
> **Countries now consuming energy at lower levels** (generally LDCs). EC/C doubles in 2020 and again in 2050 if 1988 EC/C is equal to or less than 20 and TFR is less than 5. Under all other conditions, EC/C increases by 80 percent. After 2020, LDCs are held to a level no greater than 100 EC/C.

3. MDC Energy Constraints Scenario with Lower LDC Population Growth. This scenario assumes the MDC energy constraints in Scenario 2; it also assumes that population growth in LDCs is reduced 20 percent more than already assumed in each 30-year period in Scenario 2.

4. Tougher MDC Energy Constraints Scenario. There are no initiatives today to reduce per capita energy use per se. Rather, there are proposals to reduce the deleterious effects of present-day commercial energy technology on the environment, notably the world conferences concerned with reduction of CO_2 and other greenhouse gas emissions associated with fossil fuels. Therefore, this scenario is based on the recent Second World Climate Conference, where a number of developed nations proposed taking immediate steps to cut greenhouse gas emissions. Examples: Germany proposed a 30 percent cut within 15 years [includes the benefits of closing (previously) East German plants that are inefficient and high greenhouse gas emitters]; Denmark plans to reduce its greenhouse gas emissions by 20 percent by 2005 and by 50 percent by 2030; and the members of the European Community pledged to reduce greenhouse gas emissions by 20 percent by 2005. So far, the parliaments of four of these countries (Germany, Italy, the Netherlands, and the United Kingdom) have ratified the agreement.

Since major changes in the means by which energy is produced will take many years, achieving reductions in greenhouse gas emissions will require lowering energy use in the developed world, at least in the near term. For the developing world, which has a very low level of energy consumption per capita, it is unreasonable to expect any reductions, and indeed growth must occur. For these countries, the assumptions are little different from those in previous scenarios.

Countries at or above 110 EC/C (present-day level in Japan; those at or above this level generally are MDCs and OPEC). For the year 2020, cut EC/C by 30 percent if it exceeds 200 today; by 20 percent if consumption is between 150 and 200; and by 10 percent if consumption is between 100 and 150. For the year 2050 for all countries with an EC/C in 2020 of 90 to 130, cut by 10 percent; 130 to 180, cut by 20 percent; and greater than 180, cut by 30 percent; with a floor of 90 (that is, no country need consume less than 90 gigajoules per capita by 2050).

Countries now consuming energy at lower levels (generally LDCs). EC/C in 2020 and again in 2050 doubles if 1988 EC/C is equal to or less than 20 and TFR is less than 5. Under all other conditions, EC/C increases by 80 percent. For 2020, LDCs are held to a level no greater than 100 EC/C (Japan's 2020 target) and for 2050 to a level of 90 EC/C (Japan's 2050 target).

5. Tougher MDC Energy Constraints Scenario with Lower LDC Population Growth. This scenario assumes the MDC energy constraints in Scenario 4 plus population growth in LDCs is reduced 20 percent more than already assumed in each 30-year period in Scenario 4.

RESULTS

In the next 30 years, global commercial energy consumption is projected to more than double under unconstrained assumptions. Under the toughest assumptions (most severe MDC energy constraints and reduced LDC population growth), total global energy consumption would increase by one-third (figure 2). By 2050, even under the toughest assumptions, commercial energy consumption is projected to increase 82 percent from the 1988 level. What will be the relative contribution of population growth and changes in energy per capita to global energy consumption? How much will MDCs and LDCs contribute to global totals? The remaining analysis begins with the contribution of changes in population growth, per capita energy consumption, and their interaction to MDC and LDC energy consumption totals. The same analysis is then applied to world aggregate data and the result is compared to the summed MDC and LDC data.

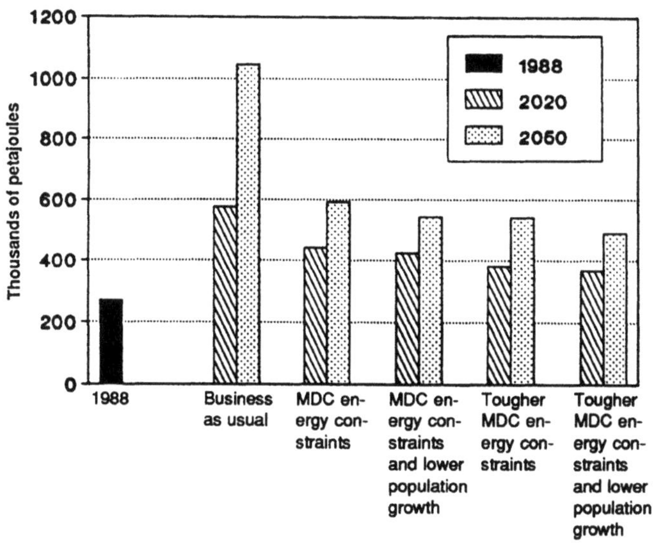

Figure 2. Global Consumption of Commercial Energy: 1988, 2020, and 2050.

The method for calculating the components of change is as follows:

-- population growth from time 1 to time 2 multiplied by energy per capita at time 1 equals the contribution of population growth.

-- the change in energy per capita from time 1 to time 2 multiplied by the population at time 1 equals the contribution of the change in energy per capita.

-- population growth from time 1 to time 2 multiplied by the change in energy per capita from time 1 to time 2 equals their interaction.

These are the values graphed in figures 3 through 6.

COMPONENTS OF CHANGE IN COMMERCIAL ENERGY CONSUMPTION FOR MDCs AND LDCs

1988-2020. The following observations can be made about the contributions of changes in population growth and in energy consumption per capita to commercial energy consumption between 1988 and 2020, as illustrated in figure 3.

-- Under the assumptions of the unconstrained scenario, the increase in commercial energy consumption is greatest in the MDCs, and that increase is dominated by the increase in energy per capita. Population growth and the increase in energy per capita are equally important factors in the increase in commercial energy use in LDCs.

-- When energy constraints are imposed on MDCs, population growth becomes the dominant factor in their increase in commercial energy consumption. With the tightest energy constraints, population growth is offset by the decrease in energy consumption per capita and total energy consumption remains almost constant.

-- Population growth in MDCs, although very low, contributes as much to global energy change from 1988 to 2020 as the much larger population growth occurring in LDCs. This is because energy consumption per capita is much higher in MDCs than in LDCs.

-- Under all scenarios, the components of change are remarkably similar in LDCs. This is because the major differences affecting LDCs among the scenarios is the reduced population growth in scenarios 3 and 5 and this has relatively little effect on commercial energy consumption in countries where energy consumption is very low. It should be noted that the 20 percent reduction in population growth on top of the already assumed reductions used in these two scenarios is very stringent.

-- Only in the unconstrained scenario is the change in total energy use (the sum of the three bars) larger in the MDCs than in the LDCs.

2020-2050. In the 2020 to 2050 period, patterns seen in 1988 to 2020 become more prominent, as shown in figure 4.

-- In scenarios 2 through 5, energy constraints more than offset population changes in the MDCs; in the unconstrained scenario, which has no constraints, increases in energy use per capita dominate population growth.

-- In the LDCs, population growth is an important factor in the components of change in commercial energy consumption under the assumptions of all scenarios, but change in energy per capita is the most important factor. A 20 percent reduction of population growth has relatively little effect on the change in commercial energy use.

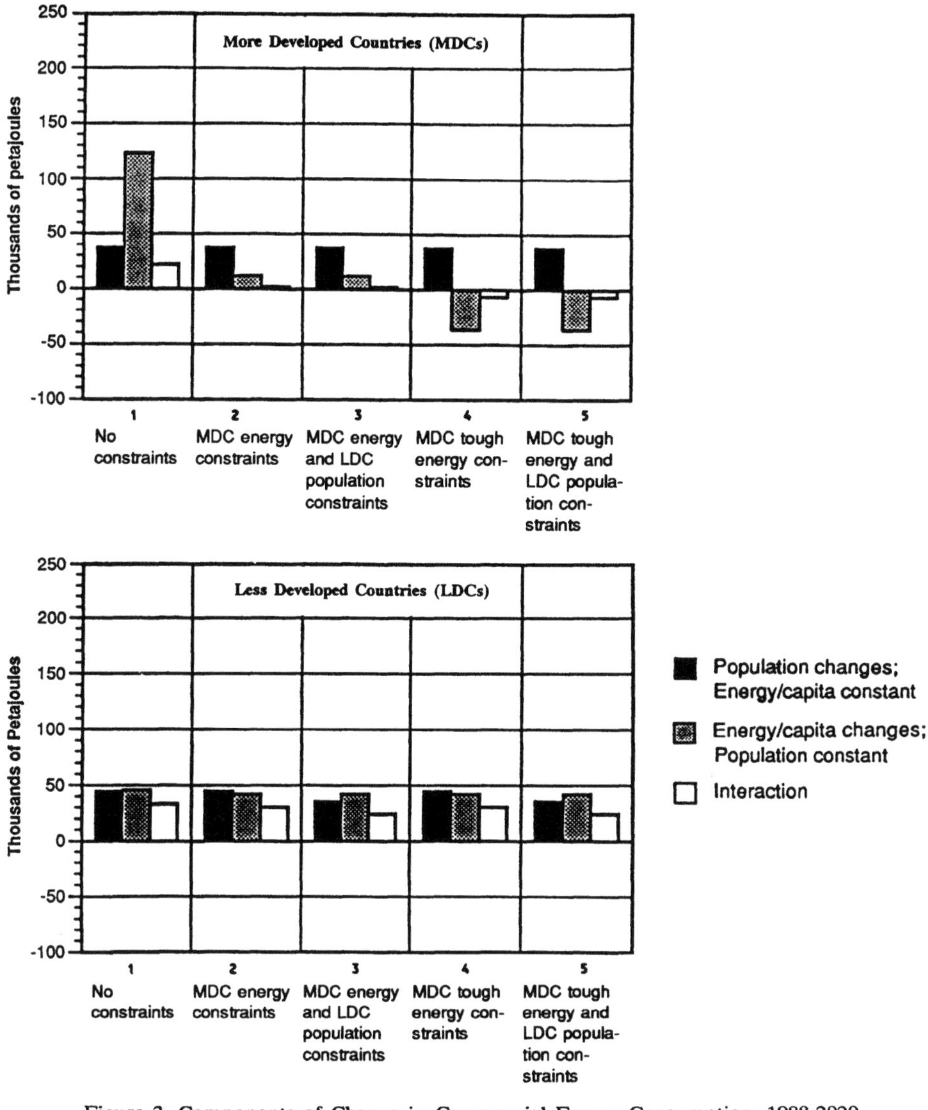

Figure 3. Components of Change in Commercial Energy Consumption: 1988-2020.

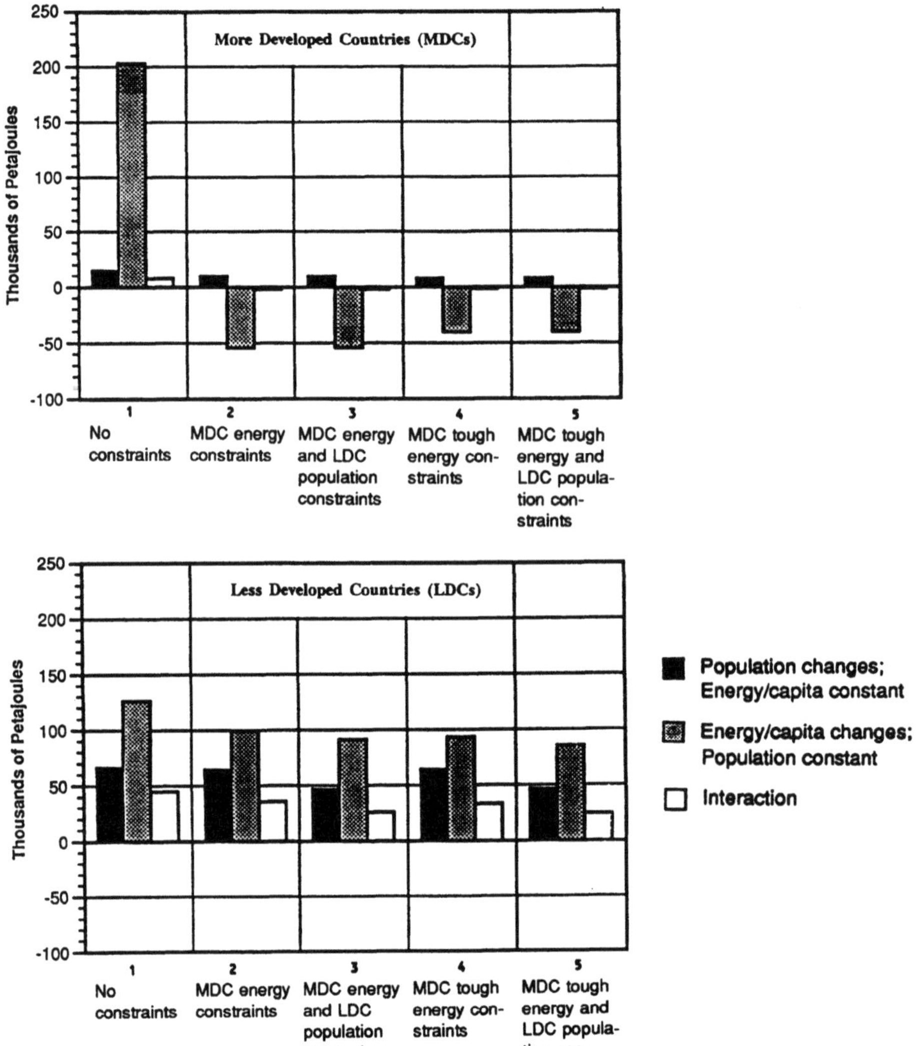

Figure 4. Components of Change in Commercial Energy Consumption: 2020-2050.

As LDCs energy consumption per capita increases, population growth becomes increasingly important in determining their energy consumption. Even if their population growth rates are slowed dramatically, the combined effect of their increased energy consumption habits with their continued population growth will make their contribution to the increase in total energy consumption larger than in the MDCs.

COMPONENTS OF CHANGE IN WORLD COMMERCIAL ENERGY CONSUMPTION

The sum of the averages is not the same as the average of the sums. If global population and energy consumption totals are used to analyze components in worldwide commercial energy use, as illustrated in figures 5 and 6, a very different picture emerges than would be expected from the separate analyses of MDCs and LDCs discussed above. Increases in population are far more important in determining commercial consumption in every scenario between 1988 to 2020 and in all but the business as usual scenario between 2020 to 2050. The reason is that very large population increases in LDCs are being multiplied by large values of energy per capita in MDCs. This kind of problem is likely to occur when aggregate data are used to describe a whole that is made up of very dissimilar parts.

A more accurate way to analyze the components of the global changes is to sum the results of the dissimilar parts as has been done to produce the bottom graphs in figures 5 and 6. The bottom graph in figure 5 is the arithmetic sum of the two graphs in figure 3, while the bottom graph in figure 6 is the arithmetic sum of the two graphs in figure 4.

Under the unconstrained scenario, change in energy per capita is more than twice as important as population growth in changing world commercial energy consumption in both periods. When energy use is constrained, as in all other scenarios, population growth increases in relative importance.

CONCLUSIONS

This paper is an exercise to better understand the relative importance of population growth in the future global consumption of commercial energy. The tentative conclusions are:

- Even with severe constraints on MDC energy use and LDC population growth, worldwide commercial energy consumption will continue to rise. Under the toughest assumptions used in this paper, worldwide commercial energy consumption grew 82 percent by 2050.

- Population growth in MDCs, although now low, contributes as much to global energy change from 1988 to 2020 as the much larger population growth in LDCs because of much greater energy consumption per capita. Between 2020 and 2050, population growth in LDCs is over three times more important than population in MDCs.

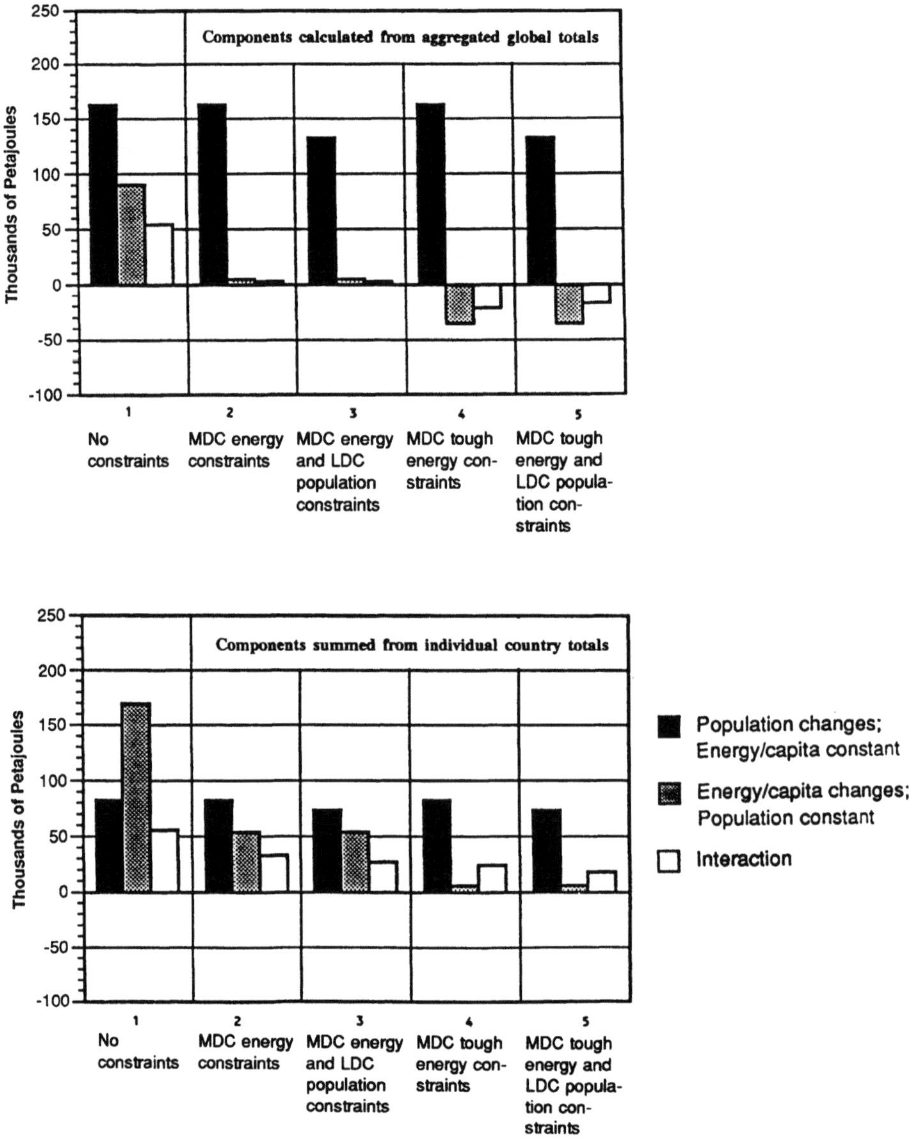

Figure 5. Components of Change in Commercial Energy Consumption, for the World: 1988-2020.

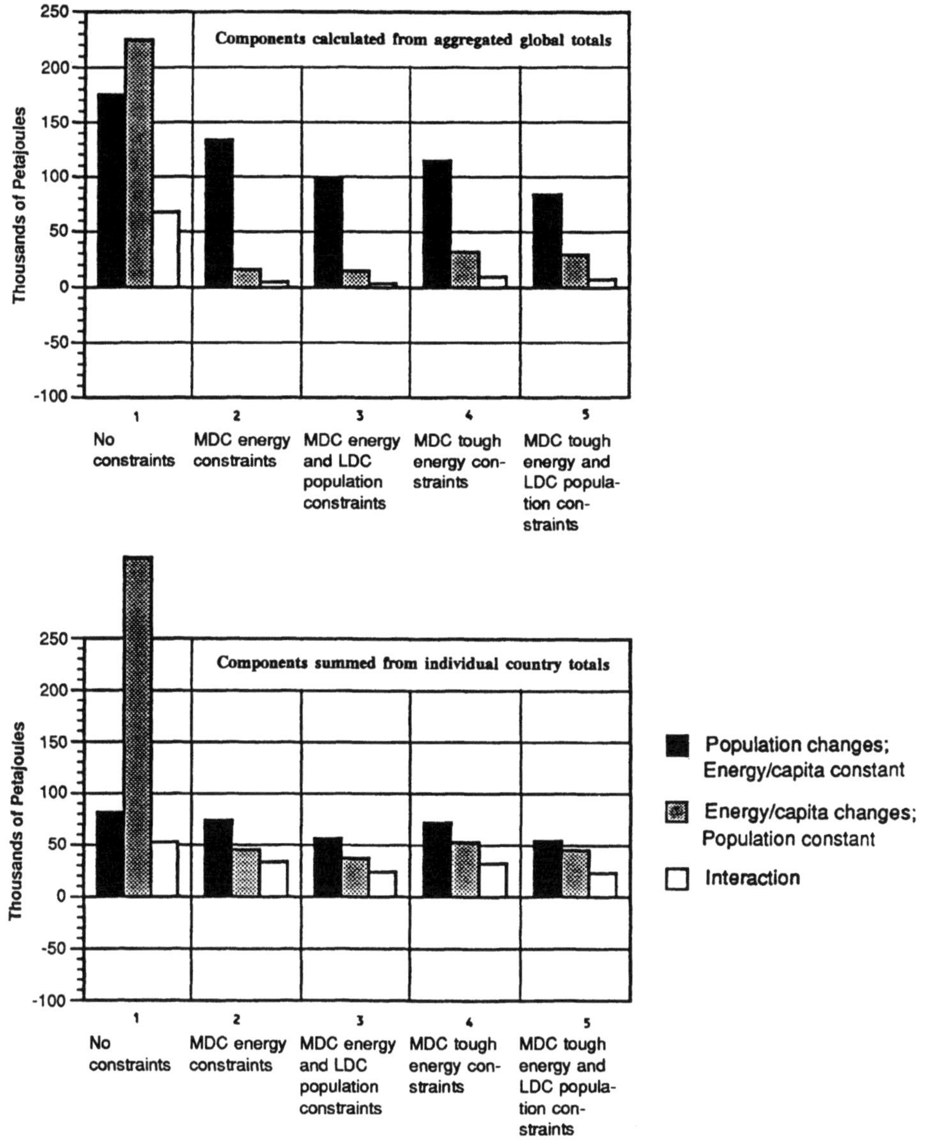

Figure 6. Components of Change in Commercial Energy Consumption, for the World: 2020-2050.

- -- While constraints on LDC population growth contribute to reduction in global commercial energy growth, such constraints are, in general, less important in the near-term than constraints on MDC per capita energy consumption under the assumptions made in this paper.
- -- Summing data from individual countries to provide global totals provides a more accurate picture of the relative contributions of population growth and per capita commercial energy consumption to total commercial energy consumption than using aggregated global data.

With respect to any implications that might be drawn for greenhouse gas emissions from this paper, it should be noted that the calculations have been done for only two points in the future and, thus, completely ignore the cumulative effects of these emissions because of the long lifetimes of the various molecules involved.

REFERENCES

Ehrlich, Paul and Anne Ehrlich. 1990. *The Population Explosion*. Simon and Schuster: New York.

International Institute for Applied Systems Analysis (IIASA). 1989. *Scenarios of Socioeconomic Development for Studies of Global Change: A Critical Review*, F. Toth, E. Hizsnyik, and W. Clark (eds.). Research Report-89-4. Laxenburg, Austria. June 1989.

International Institute for Environment and Development (IIED) and World Resources Institute (WRI). 1987. *World Resources 1987*. Basic Books, Inc.: New York.

Keyfitz, Nathan. 1990. "Population and Development Within the Ecosphere: A Bibliographic Essay." IIASA Working Paper 90-70. Laxenburg, Austria. December 1990.

Office of Technology Assessment (OTA). 1990. "Energy Use and the U.S. Economy." OTA-BPE-57, U.S. Government Printing Office: Washington, D.C., June 1990. p. 2.

___. 1991. "Energy in Developing Countries." OTA-E-486, U.S. Government Printing Office: Washington, D.C., January 1991.

Sadik, Nafis. 1990. *The State of World Population*. United Nations Fund for Population Activities, New York.

Shaw, Paul. 1989. "Rapid Population Growth and Environmental Degradation: Ultimate versus Proximate Factors." In *Environmental Conservation*, Vol. 16, No. 3, Autumn 1989. p. 199.

U.S. Bureau of the Census. 1989. *World Population Profile: 1989*. U.S. Government Printing Office: Washington, D.C.

World Bank. 1990. *World Development Report 1990*. Oxford University Press: New York.

World Resources Institute (WRI). 1990. *World Resources, 1990-91*. Report by the World Resources Institute, Oxford University Press.

NOTE

[1]The views expressed in this paper are those of the authors and do not necessarily represent the views of the U.S. Congress Office of Technology Assessment, the Technology Assessment Board, or the U.S. Bureau of the Census, Center for International Research.

THE PUBLIC AS A DECISION MAKER

Rob Coppock

Staff Director
Policy Implications of Greenhouse Warming
National Academy of Sciences
2101 Constitution Avenue, NW
Washington, DC 20418

This paper is not based on studies performed at the National Academy of Sciences. I have been involved, over the last three years, in a study of policy implications of greenhouse warming. Of course, we dealt with some of the issues that are involved in considering the public as decision makers, but what I have to say is from my own thinking and I hope you will treat it appropriately.

In April 1991, the National Academy of Sciences issued what we believe is a landmark report on the policy implications of greenhouse warming. The chairman of that report, the Honorable Daniel J. Evans, the former U.S. Senator and Governor from Washington State, briefed members of Congress the day the report was released. That briefing was attended by seventeen members. All but two stayed until Senator Evans had answered everyone's questions. Anyone familiar with Capitol Hill knows how unusual it is for that many members to stay in one room for ninety minutes. In my mind, there can be but one explanation. These elected officials knew their constituencies were worried about climate change and they wanted to know what the country's best minds had to say about responding to the threat of greenhouse warming.

Public opinion is an important influence on government policy and climate change is no exception. But in this case, public opinion, like climate change itself, is a global phenomenon. Next June, tens of thousands of people will converge on Rio de Janeiro for the United Nations Conference on Environment and Development. In conjunction with that meeting, there is considerable interest in an international diplomatic convention on air pollution and climate change.

The United States today seems to stand alone in opposing emission caps or taxes on greenhouse gases, especially carbon dioxide. The positions in favor of such actions by other countries—notably those in western Europe—have been strongly influenced by public opinion. Ernst Weitsaecker of Germany observes that environmental protection has been the single most important public issue in that

Global Climate Change, Edited by J.C. White
Plenum Press, New York, 1992

country over the last decade. Concern emerged first over Germany's dying forests, shifting later to nitrates and dioxin and, most recently, to the global issues of climate change, ozone depletion and tropical rain forest loss.

The situation in this country is not very different. In 1988 we experienced a long, hot, disastrous summer. Jim Hanson, head of NASA's Goddard Institute for Space Studies in New York, said he was 99-percent sure that the warming of the 1980s was not due to chance variation. Greenhouse warming was propelled onto the national agenda, and even became a part of George Bush's political platform.

Since then, however, those who welcomed the attention given to the greenhouse warming problem have become deeply dismayed. The administration seems to have proclaimed itself to be the world's guardian of rationality and prudence regarding climate change. Some claim that the administration is simply out of touch with the American public. Some claim that the administration thinks the public has been misled by exaggerated and biased media reporting. They say it thinks the public does not understand the complex interactions of climate change well enough to have an informed opinion.

I do not think that is what has happened at all. I think this explanation is based on an incorrect oversimplified view of how the public influences government decision making. In my opinion, the American public has sent thoroughly mixed signals to its elected and appointed officials. The positions taken by the administration are completely consistent with what the public has told it, if you include all the messages it has received.

Let me tell you why I think this. The publics in this country and abroad are, indeed, sending messages of concern about climate change to elected officials. Concern alone, however, is not enough to determine how to respond. Unfortunately, the same public sends quite different messages about the specific actions that happen to matter a lot with regard to climate change.

Fifteen years ago, in the heyday of environmentalism, many of us recognized the wisdom of recycling, but we were pessimistic about getting the public to change its behavior. Today, newspapers, bottles and cans are picked up weekly in many American cities. Businesses separate and collect office materials for recycling. Public opinion and behavior have changed dramatically. You can read about some of the direct and indirect influences on those changes in other papers in this proceedings. My point is that individual and collective decisions of the public in the area of recycling are consistent with the concern for climate change.

There are other areas, however, where the public sends signals which are not consistent with concern for climate change. In most of these areas, the messages from the public show less willingness to sacrifice or take effective action. For example, our greenhouse warming report identified many things that can be done to improve efficiency and conservation of electricity. One of the papers in this proceedings addresses that topic.

Since most electricity in this country is generated by power plants relying on fossil fuel, reducing electricity consumption is an effective way to reduce emissions of carbon dioxide and other greenhouse gases. Some things people can do are surprisingly simple. Painting roofs white and planting urban shade trees can reduce summertime demand and need for cooling. While these actions are cheap, they also are limited in the amount of emissions reductions that can be accomplished through their use.

Replacing incandescent light bulbs with high efficiency fluorescent tubes has greater potential. These tubes are much more efficient and each one lasts about sixteen times longer than an incandescent bulb. If every household in the United

States replaced just three light bulbs, the residential electricity demand could be reduced by nearly 50 percent. They do cost something like $12. But even at that price, a household should save money in the long run on each fluorescent tube it purchases and uses. The trouble is, few people have replaced their light bulbs.

Another thing people can do is improve insulation. Of course, this is nothing new. We have been talking about this for decades. The threat of climate change, however, does suggest some new ways of thinking about insulation. For example, it is just as important to insulate to keep houses cool as it is to keep them warm. In many parts of the south, houses are woefully underinsulated and we waste a lot of energy by cooling air that escapes into the environment in the summer.

The public also increases greenhouse gas emissions directly by its choice of transportation. Use of a personal vehicle instead of mass transit dramatically increases the emissions per mile traveled. The American public has also chosen to drive cars that are larger with more emissions than people in other parts of the world and, in its choice of transit or private vehicles as in its choice of cars, the public influences the emissions of greenhouse gases.

So the public decides indirectly through the officials it elects and the programs they implement and directly through many kinds of choices every day. The choices made by the public about its own behavior, however, often do not correspond to their expressed concern about climate change. I think the administration is aware of this, and that is part of the reason the administration is exercising caution and prudence regarding action on climate change. I also think that there are things that can and should be done now and there are ways that the public, in its decision-making capacity, can make those things happen.

In this country, government is organized on different levels. I have been concentrating on the federal level which is appropriate because action at the national level can have tremendous impact. In particular, government regulations of emissions or development of more efficient technologies can make a big difference.

There are, however, many things that can be done by state governments. For example, the rules governing electrical utility pricing and profit taking could be changed in many states to encourage conservation and efficiency. California, for example, is moving ahead again in the area of transportation and is forcing the development of cars relying on alternative fuels.

At the local level, building codes and zoning ordinances could be changed in ways that directly reduce emissions. Controls and incentives relating to transportation, perhaps including control of parking places, could be implemented here.

The public influences all of these decisions to a greater or lesser degree. Take the example of recycling; it is millions of individuals and households that make recycling work. This country will reduce its greenhouse gas emissions dramatically only if millions of individuals and households decide to do the things that help. Of course, there are changes that need to be made in the rules and regulations and there are incentives that can help, especially in energy consumption.

Ultimately, however, government programs need public support. More importantly, substantially reduced emissions depend upon decisions of action by the public. In climate change, the most important decision maker is the public. Individuals, families and groups influence emissions of greenhouse gases by the action they take. Hopefully, these decisions will correspond to the concern about climate change that they express in other ways.

DISCUSSION

IMPLICATIONS FOR DECISION MAKING: SOCIETAL CONSIDERATIONS

QUESTION: How is it possible that Germany has the toughest environmental regulations in the world and yet has a very strong economy? Is there some explanation for that?

COPPOCK: That is a very good question. I would suggest that there are probably gains to be made when you constructively address environmental problems that affect the way you design and structure your economic activity or your industrial activity. This probably produces efficiency in many areas. I would also point out this similar kind of phenomena in California, which has implemented a number of environmental regulations more stringent than the rest of the country but its economy as a state has not seemed to suffer. These are, however, very complex phenomena, and they involve interactions of competitiveness and labor efficiencies and other things that are difficult to pin down to simple answers.

QUESTION: Why do we always assume that a climate change will be bad?

COPPOCK: In the study that we completed this last year, we did not make that assumption. It is clear that, even with a climate warming of a fairly significant consequence, there will be winners and losers. Not all parts of the world will find themselves in environmental, climate or weather conditions that are worse than what most of us would think they are today.

QUESTION: But that is with a certain cut-off in time. If you keep extrapolating into the future, you keep getting higher and higher temperature increases?

COPPOCK: Yes, that is correct. We stopped our analysis specifically after 50 years but the increases would continue. However, the key issue is the rate at which you can respond to those changes. There are countries, poor in resources and fiscal ability to respond, who might have a great deal of difficulty capturing the consequences of positive changes. Climate change itself is something that needs to be looked at with a degree of sophistication.

QUESTION: How much success might we expect in slowing the population explosion, given the connections between development, women's education and participation in the formal work force or, in other words, lower fertility rates?

KOLSRUD: Certainly socioeconomic development is a very important factor, and that term includes lots of different things including roles of women, education of the population and so forth. There has been a lot of argument about the degree to which this is the most important factor and the degree to which other factors, such as access to fertility planning technologies, are important.

My personal opinion is that all of these things are important. Obviously in the developed world, population growth rates went down dramatically before we had nearly the level of fertility planning technology that we have available today.

But it must be remembered that the developed world never faced the levels of population growth rate that we see today. It never exceeded around one-and-a-half percent, whereas we see very high growth rates of up to three percent in the developing world today. Also, when the developed world went through what is called the demographic transition, there was a much slower equilibration of death rates, whereas what we have done in the developing world is to reduce the death rate while not reducing the birth rate, which again accounts for this massive increase.

Yes, we need socioeconomic development. We also need access to family planning technologies in order to bring down population growth rates, in my opinion. There is very good data for this and I could point you to literature if you want to see it.

QUESTION: Can you discuss the administration's current policy on family planning?

KOLSRUD: I presume you are talking about the world population. You probably know that there are ten-year conferences on world population and, at the first one in 1974 in Bucharest, the United States led the way with other developed countries, in trying to convince the developing world that population growth was reducing their chances of socioeconomic development. Although the developing countries were having high levels of socioeconomic development, their population growth rates were so high that per capita development was remaining the same or not growing as much as it would have if they had low population growth rates. The developing world was very resistant to this because it felt that the more people that it had, the more human capital it would have. They felt that this was some plot on the part of the developed world to get them to reduce their population and slow their advancement. Within about six years, by 1980, almost all the developing world had a population planning policy. They realized that their populations were growing so fast that it was a real problem.

In 1984 we had the next world population conference in Mexico City and there was the so-called Mexico City Policy where the administration astonished the world by saying that population growth is a neutral factor in development. They, in fact, reduced funding for the United Nations Fund for Population Activities (UNFPA) and reduced funding of Planned Parenthood.

The executive director of the UNFPA claims—and I think the data show it—that the U.S., in changing its position, had a significant impact on fertility rates in the world. They started going up again.

President Bush may still be the policy maker in the 1994 conference. He has continued the previous administration's positions, although it should be noted that twenty years ago he took a rather different position, in that he was supportive of family planning legislation at that time.

The problem has become very difficult in dealing with world family planning efforts is that there has been a tendency to link abortion and family planning.

Many people would say that, if you don't approve of abortion, you can approve of family planning. But somehow, in many people's minds, they neither become alternatives nor does family planning become a solution to abortion, but the same thing as abortion. How this has happened is extraordinary, but it has very much influenced the public's view of these matters.

QUESTION: What kind of educational consciousness-raising programs would you recommend to influence those daily choices of the public that should foster climate change?

COPPOCK: Well-conceived and carefully implemented programs. As I recall, there was a question in the last session about education in schools. I was surprised when I began speaking on the greenhouse warming issue and had the opportunity to speak to a group of junior high school students. What surprised me was not their sophistication or understanding of greenhouse warming, which was not great. But they jumped from that issue to ozone depletion, which has been around for a bit longer, and I was surprised at how well they understood that mankind has been emitting some pollutants that are damaging the ozone layer and that that was a bad thing. They had a very decent idea of what bad things might happen. I think that education and information programs aimed at children are vital and terribly important.

I don't think that is the only thing there is to do. There are lots of others. I don't have a program that would cover all of the kinds of actions that might be taken or all of the things that should be done, but improving individual and group understanding of the basis—including the uncertainties—upon which we have to form our judgments, as well as the consequences of the things that we do, has to be the direction that we should go.

BEYEA: I am going to put a plug in here. We put out a publication called *The Carbon Dioxide Diet*. If any of you would like to get one from me, for $5.00 you can get a great publication which solves the whole problem for you.

To respond to your point, I purchased a lot of those light bulbs and they don't fit a lot of the sockets, so we also have a problem that we need to catch up with the size of American sockets.

COPPOCK: There were informal experiments conducted by two of the scientists on the panels that we had working on the problem. One of the experiments stuck did stick in my mind very well. The experimenter, like you, purchased fluorescent tubes and tried them in all of the sockets available in his residence. It turned out that, out of 47 sockets, the fluorescent tube would fit in four.

I should also point out that one of the qualifying sockets was rejected by the experimenter's spouse for aesthetic reasons.

QUESTION: Has there been a key difference in political leadership in other countries compared to our own and do you think that is going to change?

COPPOCK: In general terms, I don't see a great difference in the political responsiveness of the political decision makers in this country and in other countries, in the sense that they are all responsive to their constituencies. I believe that European policy and political decision makers are paying a great deal more attention to environmental problems because they have a lot more serious environmental problems in Europe than we do here. Europe is a population and market the size of the United States in a much smaller geographic area. Economic activity, agriculture, all of the human activity in Europe is much more concentrated than it is here, and the effects of human behavior on the environment and the consequences simply become more pronounced. People know that and it affects the politics.

BEYEA: I have been in the energy game now for twenty years; I am sure many of you have been also. Much of the discussion that we have had this afternoon could have occurred twenty years ago.

There are some differences. The growth projections are lower and the use of market mechanisms is much more acceptable outside of academia.

We may have made some progress, but not a lot. If we are going to deal with this problem, we have to talk more directly to each other in groups like this one. Circumstances will help to force us all, who are the actors, to talk to each other and to deal objectively with this problem.

Strategies for International Negotiations

STATUS OF FRAMEWORK CONVENTION TALKS

Daniel A. Reifsnyder

Alternate U.S. Representative
to Intergovernmental Negotiation Committee and
Director
Office of Global Change
OES/EGC Room 4329-A
US Department of State
Washington, DC 20520-7818

The scientific concern about climate change had by 1988 reached a point where the World Meteorological Organization and the UN Environment Programme jointly established the Inter-Governmental Panel on Climate Change or the IPCC. They tasked the IPCC with assessing the science and impacts of climate change and developing a range of response options. The IPCC undertook these tasks in three distinct working groups, Working Group One on science, chaired by the U.K., Working Group Two on impacts, chaired by the USSR, and Working Group Three on response strategies, chaired by the United States.

Through a frenetic series of workshops, working group meetings and plenary sessions, mostly over a one-year period, the IPCC produced its First Assessment Report in August of 1990. The report set the stage for the Second World Climate Conference in November of 1990, and ultimately for the UN General Assembly's resolution 45/212 of December 21, 1990. This established a single negotiating process for development, by an inter-governmental negotiating committee (INC), of a framework convention on climate change to be completed prior to the UN Conference on Environment and Development in June 1992 in Rio de Janeiro.

The United States hosted the first session of the INC in February 1991 in Washington. As we had feared, that session proved too early for substantive negotiating progress, but it lasted too long merely to deal with organization and procedure. Not surprisingly, the session was viewed by many as a failure, particularly because it produced an intense early skirmish between the United States and India over the terms of the discussions. In effect, India sought assurances that industrialized countries would provide new and additional financial assistance and technology on preferential, noncommercial terms as a precondition of the negotiations. Not

Global Climate Change, Edited by J.C. White
Plenum Press, New York, 1992

surprisingly, perhaps, the United States rejected these demands, making clear that the commitments, with respect to financial resources and technology that industrialized countries may make, will be consistent with the obligations countries are prepared to undertake with respect to the climate issue.

While it is true that the majority of current emissions originate in industrialized countries, it is also true that emissions of greenhouse gases are increasing dramatically in developing countries and before long will overtake those of the industrialized world. This is why the United States believes that a framework convention should be a global agreement that can win the support of the largest possible number of countries.

The skirmish that I mentioned ended, as these skirmishes often do, in a compromise in which all agreed that consideration of financial resources and technology would proceed apace with consideration of commitments. This compromise enabled the parties to establish two working groups—one on commitments, the other on mechanisms. Essentially, it is in those two working groups where all of the action is now taking place within the INC.

The second session took place in June in Geneva. Prior to and during this session, countries put forward a wide range of proposals on all aspects of the convention. The session was characterized by a far greater degree of harmony and cooperation than the first session. In effect, though, countries were putting forward their ideas, but there was little substantive debate on them. We anticipated that the third session, which took place in Nairobi from September 9-20, would be far more confrontational and could include a shoot-out between Japan and the European Community on the one hand, and the United States on the other, over targets and timetables for greenhouse gas reductions.

As you may know, Japan and the EC, as well as virtually all other members of the OECD, have adopted various targets and timetables for stabilizing greenhouse gas emissions. While there are wrinkles and caveats to each of these, most center on a goal of stabilizing emissions of greenhouse gases at 1990 levels by the year 2000. Why stabilization of emissions? Why at 1990 levels? And why by the year 2000? Because these countries firmly believe that the scientific evidence of climate change is alarming, that it dictates a precautionary approach, and that by adopting a top-down approach, industrialized countries will give a strong signal domestically and internationally to developing countries of their resolve to take effective action in the near term.

While the long-term benefits of specific actions to implement these targets and timetables may be very significant, the short-term benefits of such specific actions are virtually impossible to identify and quantify. For example, no existing global climate model has the capacity to determine what climate change impacts would be avoided by stabilizing emissions at 1990 levels by the year 2000.

As you may know, the United States virtually alone among the OECD countries has rejected and continues to reject this approach. We have instead advocated a bottom-up approach under which countries would take specific, concrete actions to limit net greenhouse gas emissions. In our view, countries should seek to agree on long-term goals that are consistent with our scientific understanding of the long-term problem we face, not short-term political imperatives.

We have advocated a comprehensive approach, one that embraces all greenhouse gases, their sources and sinks, not just carbon dioxide and not just emissions from the energy sector. In our view, only a comprehensive approach will ensure that the gains with respect to one gas or one sector are not undermined or dissipated by increased emissions of gases from other sectors—as, for example, methane leaks from natural gas pipelines can negate carbon dioxide limits in the transportation sector.

In our view as well, the framework convention must be flexible, accommodating diverse national economic and political factors and natural conditions. The convention should also be dynamic, enabling the parties to take into account the need to respond appropriately to improved scientific understanding about climate change. We note, for example, that IPCC scientists indicated that greenhouse gas emissions would need to be reduced by 60 percent from today's levels to stabilize their concentrations in the atmosphere. Clearly, current technology cannot achieve such a reduction. But if we are concerned ultimately with the prospects of dangerous anthropogenic interference with the atmosphere, should we avoid establishing a long-term goal and seeking to develop the technologies we need to achieve it?

In my estimation, the reason we did not have a shoot-out in Nairobi is because there is widespread agreement with respect to long-term goals and general obligations. Where we differ with our colleagues in other industrialized countries is with respect to the methods we're prepared to take to achieve them.

Given our high dependence on domestic coal reserves for basic energy supply, the United States simply does not have the near-term options that may be available to the European Community which, through fuel switching from highly subsidized coal to natural gas (imported alternatively from Norway, the Soviet Union or North Africa), may be able to reduce its carbon dioxide emissions dramatically in the short term at a cost significantly less than the cost that the United States would bear to achieve commensurate reductions. In this connection, I would note that coal emits significant quantities of greenhouse gases per unit of energy produced, approximately 25 percent more than oil and 70 percent more than natural gas.

The United States is not alone in its concern with respect to the cost of significant near-term reductions. Countries such as Australia, Brazil, China, India and Russia have similar concerns. Moreover, the OPEC countries are concerned with the significant effect that near-term reductions could have on oil revenues.

The climate issue gives rise to a wide variety of perspectives which do not lend themselves to facile analysis. The Pacific and Caribbean Island countries have formed a new coalition that has in fact been recognized in the negotiations, known as the Alliance of Small-Island States, or AOSIS. They are particularly concerned with the prospect of sea level rise from climate change as the world's oceans expand in response to increased global mean temperature, and they have actively promoted responses that would enable them to adapt.

Similarly, drought-prone countries of Africa have sought special recognition, as have developing countries that are heavily dependent on exports of petroleum, as have developing countries that are heavily dependent on exports of forestry products, as have the least developed countries, as have the so-called Eastern European economies in transition, and so on.

Efforts within the INC to establish various categories of countries are rooted, in my view, in recognition that national and natural circumstances vary widely, and that the framework convention could affect the livelihoods of all countries in various ways, as well as the terms of trade among them. I believe that the number of specific situations that countries face will closely approximate the total number of countries in the world. Rather than generate artificial categories, the framework convention should acknowledge these differences and provide the flexibility for all countries to do their part consistent with the diverse situations in which they find themselves.

What happened in Nairobi? In essence, Working Group One dealt at length with the issue of principles at the insistence of developing countries, particularly

China. Working Group One also dealt with commitments, both general and specific. Frankly, there was little progress made in Working Group One and wide differences remain.

Working Group Two dealt rather more successfully with less controversial issues such as scientific research, systematic observation, mechanisms for financial assistance and technology cooperation, dispute settlement and standard legal clauses. Neither group has yet produced a negotiating text, but it is possible that such a text could emerge during or after the next negotiating session in December in Geneva. In fact, I would list that as the primary goal of the session coming up in Geneva—to produce a single negotiating text, however heavily bracketed it may be.

What's next? We have two scheduled negotiating sessions between now and the UN Conference on Environment and Development in Rio in 1992. We have four to six weeks of negotiations that remain. Will we have a convention? We have every reason to think so. Do we want a convention? Clearly, yes, we do. If we are to begin dealing seriously with this global problem, we must launch the process now. But we must do so in a way that builds confidence, that is flexible and allows us to adjust and adapt over time, that acknowledges different circumstances of different countries, but that enlists the support of all countries in a truly global effort. Some may say that such an approach, if only that, lacks substance and is devoid of meaningful action. We believe that substance and meaningful action must be considered over the long term, must be commensurate with our understanding of the problem we face, and must be global and comprehensive to be effective.

EQUITY IN INTERNATIONAL AGREEMENTS

Richard Stewart

Visiting Professor of Law
Georgetown University Law Center
600 New Jersey Avenue, NW
Washington, DC 20001

The atmosphere is the classic global commons. What we are learning from science is a variation of Garrett Hardin's famous essay on the commons. He took as his example a group of herdsmen whose cattle were grazing on the commons. The incentive of each herdsman is to increase his herd. The commons is overused. Each herdsman realizes this consequence, but has no way of preventing others from gobbling it up. So the incentive of each is to use up the resource as fast as possible before it is gobbled up by others.

Now, from the lawyer's viewpoint, the solution is to establish some institutional regime that limits the use of the commons. A necessary part of such a regime is to have some system for allocating who gets to use how much, and some regime of compliance assurance. The allocation of limited use rights and how limited they should be is going to depend on science and on economics, and inevitably raises questions of equity. I am going to talk about equity a little bit, and then about efficiency and how they may relate.

The nations who are the dominant players here are going to be reluctant to agree to limit their use of the global commons, unless sooner or later everybody else who counts does so also, and the allocation of resource rights is perceived as fair. As Dan has said, there are many different interests among the different players here. They cannot be just divided up North-South. Different players are suffering the adverse effects of climate change in different ways. Obviously, the island nations are the most vulnerable. I have heard talk from Russia that maybe there will be gainers: Siberia will bloom. But even if that is not so, the losses in the near term at least are differentially allocated, and the economic and social burdens of limiting the use of the resource are also quite variable. So how in this welter of conflicting and differential interests are we to get agreement?

As I look at the principles of equity that might apply, one could draw up quite a shopping list. You might say, the atmosphere is the common heritage of all human beings, and therefore we ought to allocate the right to use the atmosphere on the basis of population. Quite apart from the concern that this allocation might reward nations with irresponsible population policies, the developed nations could come back with a classic principle of the law for allocating the commons: first in time, first in right. Under this principle, existing shares ought to be maintained. To redistribute on the basis of population would wreak havoc on established economies and expectations.

A federal system or the European Community confronts similar issues in allocating the burden of environmental controls. Sometimes we have agreements on limiting emissions based on an equal percentage reduction from some baseline. How does one establish the baseline? In other cases, the solution is adoption by all of a uniform "best available technology" level of control. One might, from a different perspective, say that the resource ought to be allocated in such a way that it is used most efficiently, that is, that nations such as Japan that get the most output out of a given unit of greenhouse gas ought to get most of it, because that way we will maximize economic productivity per use of the commons. Of course, from the viewpoint of the developing nations, you might say that they have special needs, and that the allocation should be inverse to gross national product. Each according to his need, as Karl Marx said.

If you look at the principles that have emerged in various super-national negotiations, it is hard to see any one principle of equity that is going to command universal assent, and each nation will presumably advance that notion of equity that is going to be most congruent with its interests. Over the long run, I think we cannot hope for a common principle of equity. But perhaps convergent notions of equity will arise out of the interplay of these various claims and interests. It is perhaps instructive that the European Community for a long time was struggling over controls of emissions of SO_2 and NO_x from major industrial facilities. It began with equal percentage of reductions from baseline proposals, but in the end a differential allocation of cutbacks and even increases for some of the southern member states emerged from a five-year process of negotiation. I hope that we will get some agreement on greenhouse gas limitations by such a process, though I think it is going to take a fair number of years before everybody signs on to any such allocation.

The question of equity in allocation is complicated by two other equitable concerns. One is the claim of the developing countries to some assistance from the developed countries. The model here is the Montreal Protocol for the phase-out of the CFCs, where the incremental costs of agreeing to a phase-out of CFCs were effectively borne by the developed countries. That is taken as a precedent. Developing countries say to the developed nations:

> You've used up a lot of this resource already, your greenhouse gases have been up there and they're going to be up there for quite a long time. You've gobbled up an undue share of the resource and it's only equitable that you provide some compensation to us if you ask us to limit our efforts to match you in the similar use of the resource.

A related equity issue would be claims for damages or reparations. For example, the island nations might make such a claim if indeed there is a sea level rise that can be attributed at some point to the greenhouse gas emissions. The debate and the dialogue in the negotiation process reflect these several aspects of the question of equity.

Let me now turn to efficiency. Equity and efficiency are often contrasted. But here I think they are complementary. Efficiency can aid in the search for an equitable solution. Here I refer to the comprehensive approach and to trading of resource allocations between nations. The logic of the comprehensive approach is very simple. A number of greenhouse gases contribute to the threat, and global warming and the gases are a product of both emissions and sinks. All of the factors that account for changes in net greenhouse gases ought to be taken into account in an agreement. One would construct an index that compared the global warming potential of different gases with reference to a base of CO_2. One would also use the index to factor in the destruction or enhancement of sinks, in terms of their effects on net greenhouse gas emissions.

The argument in favor of a comprehensive approach in relation to equity concerns is pretty simple. One, the comprehensive approach provides a level playing field. Limitations based only on CO_2 or only on methane or only sink preservation would benefit some nations relatively and disadvantage others. Any partial limitation will tilt the playing field, and would be selected and pushed by some nations in order to obtain a relative competitive economic advantage.

The second justification for a comprehensive approach is lower cost. If nations have the flexibility, some allocation based on a net greenhouse gas index allowance to select whatever mix of measures they want—whether it be CO_2 emissions, methane emissions, or enhancement of sinks to achieve that goal—is going to be less costly than a commitment that says everybody must control CO_2 emissions. The cost of achieving stabilization of net greenhouse gases and maintaining it will be large. Use of a comprehensive approach to reduce the economic and social burdens of compliance will promote the chances of agreement.

Practical problems have been raised about constructing an index and implementing a comprehensive approach. The dramatic re-evaluation, in light of recent science of the contribution of CFCs to global warming, shows how uncertain the scientific basis for policy choices may be. We had thought that the CFCs in the last decade were contributing a quarter or more of all the increases in greenhouse gases, and now the recent science says no, on net they are not contributing anything. Our understanding may change again tomorrow.

But even a single gas or a single substance approach to regulation may face the same problem. For example, we have a massive regulatory program in the United States, with very stringent limits on various industrial sources, to control dioxin. Our evaluation of the threat of dioxin is being dramatically revised downward. In the global warming context, CO_2 is by far the dominant part of the problem, but its exact contribution is plagued by uncertainties over residence time and missing sinks. Nonetheless, the virtues of a comprehensive approach remain. I think there will be adequate science to warrant a CO_2 equivalent credit for reducing methane or preserving forests. The way to push the science along and establish credibility is to have a CO_2-based index and put the burden on those who want to get credit for something other than CO_2 to establish the requisite degree of equivalence. At least for some gases like methane, enough is known to do that.

In addition to the comprehensive approach, we ought to have international trading of net greenhouse gas emissions reduction credits. We could, for example, have country-to-country trading. If Brazil has a negotiated allowance of net greenhouse gas emissions, and the United States or Japan proposes to help Brazil build factories to produce energy-efficient refrigerators or automobiles, Brazil might provide them part compensation for the transfer of capital and technology by transferring some of its greenhouse gas emission rights that it will not now need because of the increased energy efficiency. Or the trading approach can work on a private firm-to-private firm basis. A Japanese firm investing in such a plant in Brazil could be compensated by the plant for the firm's transfer of capital and technology with greenhouse gas emission rights which the Japanese firm could sell to another firm in Europe who wants to build a new plant. We would have a worldwide green currency. Some international agency would have to do the accounting and make sure people are not cheating by emitting more greenhouse gases than they have allowances for. Trading would further reduce the costs of limiting global warming by ensuring that capital and technology is channeled any place in the world where it is cheapest to reduce net greenhouse gas emissions. It would achieve the greening of Adam Smith and David Ricardo.

The other virtue of a trading approach is that it provides a cost-effective and decentralized way of transferring capital and technology from the developed to the developing nations. Increasingly the problem of greenhouse gas emissions will be in the developing nations. I am skeptical that there is going to be the political will in the developing nations to transfer, on a bilateral or multilateral basis, the amounts of capital and technology that will be needed. The more indirect transfer through trading can close a lot of the gap.

Trading intensifies what is at stake in the negotiations, because these emission rights are a form of capital, an economic resource. But I think it is all the more appropriate that we make clear what is at stake, and ensure that rights to emit net greenhouse gases are used most efficiently. I hope that the Administration will continue to push the comprehensive approach and ensure that there is freedom for consenting adult nations, at least, to trade their emission rights. Over the long run, given the magnitude of what is at stake, we need to do this job in the most rational and cost-effective way that we can.

DISCUSSION

STRATEGIES FOR INTERNATIONAL NEGOTIATIONS

QUESTION: Given that science tells us that the more greenhouse gases are pumped into the atmosphere, the sooner their effects will be felt and the greater those effects will be, isn't it entirely possible that the U.S. no-regrets policy will turn out to be a deep-regrets policy?

REIFSNYDER: One point that I didn't make in my presentation is what this negotiation is all about. What kind of an insurance policy are we willing to buy collectively with respect to the issue of climate change? I think that most countries feel that the potential for climate change is real. There is a lot of disagreement about how quickly and how severe and what the science tells us and so forth, but I think overall there is a great deal of concern, and there are some uncertainties involved. So the question isn't about doing nothing; the question is what we do and how much we do. In a very real sense, it's how do we deal with all of these uncertainties and difficulties and the differences among countries and so forth? What is the best insurance policy we can take out, and what is the premium that we're willing to pay at this time?

One of the analogies I've heard is to life insurance. You take out a life insurance policy to guard against certain eventualities, but by the same token, you want to make sure that the premium you're paying isn't going to destroy you in the process. The issue of trying to decide what that premium is going to be is where we are, and our perception of what it's worth may change over time. I think we're going to start relatively modestly for the very real reason that, for example, the Western world and certainly the United States is in a recession. Funds are limited, and I think policymakers are going to have a very difficult time trying to convince people, based on our knowledge at the moment, that deep economic sacrifices anyplace are in order.

QUESTION: What role might the availability of abundant natural gas in the USSR have on the enthusiasm shown in Western Europe, especially Germany, in reducing CO_2, considering the lower CO_2 output of gas versus coal, and the willingness of the USSR to sell its gas for hard currency?

REIFSNYDER: My answer is, a lot.

QUESTION: Yesterday we heard support for a technology cooperation regime emerging from industry presentations. In what form, if any, do you predict such a regime will be established under the Climate Convention, in light of the fact that many developing countries have a need for rugged rather than state-of-the-art emissions technology?

REIFSNYDER: Frankly, technology cooperation is at the heart of the issue with respect to developing countries. I believe that we have put forward a number of proposals related to technology cooperation and, in fact, I think at this next session of IPCC we will have yet further evidence of what the United States is doing. We are doing a great deal in the world right now with respect to this. Ed Hillsman from Oak Ridge, who is in the audience, has helped us produce a document that we will be distributing at the climate negotiations in Geneva next week.

I think focusing on the kinds of technologies that developing countries need in their particular situations is going to be key. This is why we believe that some of the rhetoric we hear about preferential noncommercial access to technology and so forth is very unfortunate, because a great deal of what we believe is needed already exists. The questions are how to disseminate information and these technologies to the countries that need them, and how to determine the kinds of technologies that they need. There is a great deal that can be done, and I think the entire negotiating process increasingly is turning on the issue of technology cooperation and how best to ensure it. I think there is a willingness on the part of industrialized countries to get deeply involved in this with developing countries. There is certainly a need and interest in this on the part of developing countries.

QUESTION: There has been reassessment of the role of CFCs and the fact that, therefore, the net emissions of greenhouse gases in the 1980s are less than we thought. How does this influence IPCC negotiations?

REIFSNYDER: I was with Bill Nitze the day after the ozone assessment came out. Bill said two things about it. First, there's good news and there's bad news. The bad news is that things are worse with respect to the ozone layer than we expected. The good news is that we may have more time in dealing with climate.

I'm not sure that, in the dynamics of the negotiations, we perceive we have more time in dealing with climate. I don't think it's going to cause the negotiating parties to throttle back. On the other hand, people have said, if CFCs are no longer a part of the problem, then clearly CO_2 is much greater. I think that may be true, but so is methane, so is nitrous oxide. In fact, as Bill was saying, you have a pie chart that once had a 26-percent share occupied by CFCs. That 26-percent share is distributed now to other gases, but by the same token, the size of the pie shrinks.

QUESTION: How are we going to calculate an index, or how will one be developed, and can consensus be reached?

STEWART: The progress on the scientific parts of the IPCC and its follow-ons have been very encouraging. It has been shown how the international scientific community can engage in the dialogue, can move on the state of science. The United States administration has given a lot of priority to that science; it's spending a million dollars a year more on it. We're taking up about half of the world-wide science in terms of the resources that are being devoted to it. I think the experience with the Montreal Protocol and its follow-on suggests that the international science community can make progress and can reach a consensus about those issues.

QUESTION: Many countries may use regulation internally to control emissions. Will that make trading more difficult?

STEWART: I don't think analytically that it necessarily does. You can have trading between countries and yet each country internally can use either a tradable market system or a regulation or socialism or whatever. I believe that market-based incentive approaches are the preferable way, not only between nations, but internally, and that some form of international trading can set a precedent for trading internally. I think there are going to be opportunities to do the experiment with trading on a regional level. One very promising opportunity on the horizon is a follow-on to the U.S.-Canada agreement on acid rain—to have transborder trading in sulfur rights, to extend the trading scheme under the Clean Air Act across the border.

I think eventually in Europe, within the Community, we are going to see some form of cross-country trading on SO_2 and NO_x. I'm not suggesting a trading system be imposed from the top down; it's going to grow actually from the bottom up.

QUESTION: What mechanisms or sanctions would exist to prevent a nation from exceeding its allocation of greenhouse gases?

STEWART: I suppose it's going to be, at least in the near- to medium-term, publicity, accounting, the scrutiny of the world community and side sanctions. If you're flagrantly violating your commitments, you're not going to get favorable trade treatment. We're not going to have an international police force. I think it would be a mistake to set up some international court in the near-term to adjudicate disputes. It's going to be by much more informal processes.

QUESTION: Wouldn't a tax scheme be preferable, instead of trading, in order to raise the revenues which could derive funds for assistance to the developing countries?

STEWART: Of course, under a trading comprehensive approach, a nation internally could use taxes to drive reductions, and the European Community is talking seriously about CO_2 taxes. I think internationally, logically, one could say that's an alternative approach. The French, at some point, proposed it with a carbon tax that was a graduated tax. Interestingly, the progressive element of the tax started to go up sharply at levels of carbon use just above that of the French economy.

I think it's unlikely that nations are going to hand over the amount of revenues we're talking about, which would be very considerable, to some international bureaucracy. I'm not very confident that an international bureaucracy would spend the money very well or in the most efficient way in promoting the technologies and changes needed in the developing countries.

QUESTION: Given that the results of permit trading have proven mixed for lead and phosphorus, won't there be more potential for failure in alternative markets?

STEWART: I don't know about the phosphorus. The lead phase-down was quite a successful system. What it did was greatly ease the cost and provide the flexibility for a quite dramatic reduction in lead gasoline, by allowing different firms to phase different costs or different time horizons in coming into compliance to trade.

QUESTION: Who would police complex schemes such as greenhouse gas trading?

STEWART: I think one would need some sort of international Securities and Exchange Commission, some sort of international registry. It is a more complicated scheme than an across-the-board reduction of emissions by X. I think, institutionally, the question is whether the gains in terms of flexibility and efficiency and the dynamic incentives that are provided are sufficient to justify the added administrative costs. At least, with the lead phase-down as an example, the answer is clearly yes, but that was a rather limited experiment.

We're going to have an experiment on a much greater scale with the sulfur trading provisions in the Clean Air Act, the acid rain provisions, which will provide a case study. And we're going to have opportunities for regional experimentation,

U.S., Canada, and in the Community. So we'll get some experience on this. I'm rather optimistic. I think the evidence to date justifies that optimism, but we're going to have to experiment with the approach.

QUESTION: Will the Bush Administration acknowledge that there is indeed the threat of global climate change?

NITZE: I think I can answer that question unambiguously, yes, with one caveat. On January 30, 1989, Secretary Baker, in his introductory remarks to the policy working group under the IPCC here in Washington, laid out the so-called no-regret strategy, namely, that we should not wait for further scientific knowledge or economic knowledge about costs and benefits before beginning to act, and that we should now take actions—such as phasing out CFCs, improving energy efficiency and reversing deforestation—that made sense for other reasons. To my knowledge, the Administration has never repudiated or significantly modified that statement of policy by the Secretary of State. Inherent in that statement is a recognition that there is at least a risk to the international community from global climate change.

Secondly, at the start of the intergovernmental negotiating committee process in Chantilly last February, the Administration presented a document entitled, *America's Climate Strategy*. Normally the government of a country like the United States does not put out a strategy to address a problem unless it believes that the problem really exists.

Thirdly, although the Administration regrettably hasn't received credit for it, the Administration has taken the lead on an international program of research on global change which not only includes scientific research, but also includes research on the impacts of change and policy responses. I think that initiative is important and is making headway. We are in the process of negotiating with the Mexicans and the Brazilians and other Latin American governments for the creation of a major center of policy research. This would be the center of a network throughout the world to really look at these issues and address them.

The caveat is this ideological resistance to admitting that there is an urgent problem because of the policy implications of such an admission, policy implications that might lead to imposing economic costs on politically powerful constituencies like the coal industry or offend conservatives within the Republican Party or cause some electoral difficulties for the President in the '92 election. But that's a caveat. Intellectually, the Administration has admitted that there is a risk, that this is a problem and that it needs to be addressed.

QUESTION: What is the role of the World Bank, particularly on energy development, and what is the role of corporations?

NITZE: The World Bank's energy policies have in the past been, in light of what we know today, somewhat misguided. But the World Bank is in the process of policy reform. That is not going as quickly as I would like to see, but I would predict that there won't be major World Bank funded coal projects without very careful examination of the environmental costs and benefits, that there will be increased World Bank spending on efficiency in renewables, that the World Bank will become more seriously involved in working with major developing countries like India and China and Brazil in developing a very different kind of energy strategy and a very different kind of energy and forestry strategy than those that have been advanced in the past. I am an optimist about the commitment of the new president of the World Bank and his associates to change World Bank policy in this area and to become a real partner in sustainable development.

QUESTION: Why we are focusing so much on governments? Isn't what corporations do more important?

NITZE: In the United States that's true. There is some good news. Major American corporations have indeed taken the lead on specific actions in line with the Administration's no-regrets policy to reduce greenhouse gas emissions. In the electric utility industry, three utilities with which we work closely at the Alliance—the Pacific Gas and Electric Company, the Southern California Edison Company and the New England Electric System—have announced, in effect, greenhouse targets. Basically what they've said is that, in California and in a large part of New England, greenhouse gas emissions from the electricity sector will go down, not up, in the 1990s, and that they will get almost all of their new electricity supplies from a combination of efficiency improvements in renewables and will not build many, if any, new fossil-fuel fired stations.

In New England there is a large switch from oil to natural gas, which is a major part of this program, so there will be the construction of more high-speed gas turbines. But in any event, there is a major shift going on within the utility industry, particularly on the coast, but increasingly in the middle part of the country and even in the hinterlands. Georgia Power, for the first time, is really working seriously on demand-side management, which is a real turnaround.

There are a number of other examples. The waste management program at 3M is famous. Dow Chemical has made fantastic advances in terms of improving energy efficiency and reducing emissions. In the oil industry, Unical has had a very successful experiment in Los Angeles in retiring very dirty and inefficient cars through a junker bounty program, paying $700 a car. Cumulatively, all of this activity by the corporate community will make a difference.

Furthermore, corporations are beginning to organize and establish a world-wide network through the Business Council for Sustainable Development and within the International Chamber of Commerce. There are a whole variety of international activities within industries and among industries which are beginning to have a major impact. So, in a sense, I regard the corporate communities as being ahead of the government in terms of a practical cost-effective response to this problem.

REIFSNYDER: I think that Governor Sununu's resignation probably will prove to be a problem for us in the next round of negotiations. The Europeans, many of whom have long felt that our policy is dictated by Governor Sununu, will see that with his absence the way will be open to turn up the heat on the United States and get us to buy into the target-timetable approach to limiting greenhouse gas emissions. So I have no illusions that this is good news from our standpoint for the negotiations, if only because I think that the going is going to get tougher.

I think it was bound to get tougher the closer we get to the so-called end of the negotiations, at least on the framework convention. On the other hand, I would also echo something that Dr. Nitze said. I have no reason to think that the United States policy is going to change at all with respect to the approach we've taken, because of Governor Sununu's resignation. I think that the policy is not rooted in a single individual or a single set of views. I think it's based on many factors, and it's a widely held view in the Administration by any number of people.

QUESTION: Is it possible that the U.S. would support convention language which calls for stabilization of greenhouse gases at 1990 levels by 2000?

REIFSNYDER: If that stabilization could be accomplished by allowing nations to consider all greenhouse gases and take those actions first which will be of no-regrets, i.e., the ones without adverse economic impact.

I could start to answer that in reverse and work my way back to the primary question. First of all, I do believe that the convention will take a comprehensive approach, that it will look at all greenhouse gases, their sources and sinks. As Dr.

Nitze said, I think Dick Stewart has certainly had a great deal to do with pushing that approach. We have made tremendous progress internationally in convincing the rest of the world, and particularly developing countries, that this is something that is very much in the interests of all countries. So I believe that any convention we have is going to be a comprehensive convention in that sense, that it will embrace all greenhouse gases.

With respect to taking actions first which will be of no-regrets, I think that economic reality simply dictates that you're going to be doing those things first which make sense for a variety of reasons, rather than things which perhaps only respond to climate, simply because there are probably so many of them. It's only sensible to take those kinds of actions first.

Is it possible that the U.S. would support a convention which calls for stabilization? I have to say no, I don't think so. We have long opposed that approach, and I see nothing on the horizon that gives me any reason to think it's going to change.

QUESTION: Bill Nitze says we can stabilize greenhouse gas emissions by 2000. What's the problem with making such a commitment?

REIFSNYDER: What kind of a commitment is that? In many ways, it is a political commitment. It is a political action designed to, as the Europeans say, give a signal domestically and give a signal internationally. In that sense, how different is it from any other signal that you might give? Why stabilization, why 2000, why 1990 levels? There is no scientific basis that we know of for that particular kind of political commitment. That is a kind of a top-down command control approach to the problem which we think is fundamentally flawed.

QUESTION: What are the chances that President Bush will attend the Rio summit?

REIFSNYDER: I don't know. The President has run into unfortunate criticism recently with respect to his actions and travels abroad. That kind of criticism, if it were to continue, could well have an influence on his decision with respect to Rio, and I think that would be most unfortunate.

The Ethics of Living in a Global Greenhouse

CORPORATE AND PERSONAL RESPONSIBILITY

Dale Jamieson

Professor
Center for Values and Social Policy
University of Colorado
Boulder, CO 80309

When we look at programs like the U.S. Global Change Program (Committee on Earth Sciences, 1991), it is very easy for us to lose sight of the fact that the climate change problem is to a great extent a behavioral problem. Although most of our research dollars are spent on addressing the physical science dimensions of climate change, the problem that we face is largely a consequence of the behavior of Europeans and Americans over the last two centuries. As the developing countries industrialize, the causes of this problem will increasingly become democratized.

When we do focus on the behavioral dimensions of the climate change problem, too often we focus exclusively on how to structure economic incentives in a way that contributes to climate stabilization. The focus is on economic incentives because many people believe that human action is fundamentally and always self-interested. On this view the only lever we have on human action involves the manipulation of prices.

In no way do I wish to slight the important economic dimensions of the climate change problem. Significant work is being done in this area. However, it is very important to realize that people's value commitments and moral convictions can also figure in producing their behavior. We cut ourselves off from an important way of understanding and motivating human action when we focus exclusively on economic incentives.

This is a lesson that should have been learned from studies of the adoption of energy-efficient technologies. Utility companies that offer the same financial incentives for home retrofits often have participation rates that vary by as much as ten times (Stern et al., 1986).

The value dimensions of social problems are well understood in electoral politics. Although people may, for rhetorical purposes, compare our economic welfare today with what it was four years ago, I think that most successful politicians realize that people's attitudes towards their present economic position is at least as important as the actual economic facts about their welfare. If we think about some

Global Climate Change, Edited by J.C. White
Plenum Press, New York, 1992

important recent government actions, for example the Gulf War or the War on Drugs, I think that it is clear that people's support for these activities largely flowed from their commitments to certain moral ideals rather than a sense of their own economic self-interest.

There has been interesting work that bears on these questions. During the Vietnam War, for example, general foreign policy views were much more important predictors of people's attitudes towards the war than any direct experience with military service (Lau et al., 1978). On the issue of health insurance there is no significant correlation between people's attitudes towards the seriousness of the problem of access to health care and whether or not they personally are insured. In the case of unemployment, the fact of unemployment appears to have less influence on people's political behavior than their attitudes towards unemployment (Orren, 1990).

There is resistance to talking about ethical and value issues for a number of reasons. One reason is because many people think that focusing on ethics and values is excessively individualistic, and what we need with respect to environmental problems are collective and institutional solutions. This reservation overlooks the fact that our values permeate our institutions and practices. Changing our values is part of constructing new moral, political and legal concepts. A second reason for resistance is that many think that discussion about values is inherently idealistic. "You can't change human nature," it is often said. But I think that if we pay serious attention to disciplines such as anthropology and history, what we will see is that our current values are at least in part historically constructed. Indeed, they are rooted in the conditions of life in which those values developed. What we need to approach the problem of global climate change is new values that reflect the interconnectedness of life on a dense, interconnected planet.

There are a number of fruitful ways of thinking about the climate stabilization problem. One way of thinking about it is as the world's most difficult and complex collective action problem, both among and within countries. There are different points of view, perceptions, beliefs and values among governments and also within single nations. Sectors divide against each other in important and interesting ways.

Generally collective action problems raise concerns about assurance and compliance. Coercion sometimes answers to these concerns but it can be expensive and uncertain. Coercion is especially problematical when a problem exists on a global scale, and when there is serious diversity of perceptions and interests. One important benefit of seeing the climate change problem as involving a moral dimension as well as economic and scientific dimensions, is that it brings this problem into the domain of dialogue, discussion and participation. Rather than being solely a management problem that governments or experts can solve for us, when seen as an ethical problem it becomes a problem for everyone to address both as political actors and as everyday moral agents.

Those who seek quick fixes may find this frustrating. A moral argument does not change the world overnight. Collective moral change is fundamentally cooperative rather than coercive. No one will fall over mortally wounded in the face of an argument. Nevertheless, I think that this kind of slow change and consensus building is crucial to building a new order that will enable us to live sustainably in the twenty-first century.

What I hope to have done by this point is to have given you some reasons for taking the value dimensions of the climate change problem seriously. Now what I want to do is to say in a little more detail what I mean by values.[1]

A system of values, in the sense in which I use this term, specifies permissions, norms, duties and obligations. It assigns blame, praise and responsibility, and it provides an account of what is valuable and what is not. A system of values provides a standard for assessing our behavior and that of others. It also provides a measure of acceptability for government action and regulation.

Values are more objective than mere preferences. A value has force for a range of people who are similarly situated. A preference may have force only for the individual whose preference it is. Whether or not someone should embrace a particular value depends on reasons and arguments. We can rationally discuss values, while preferences may be rooted simply in desire without supporting reasons.

A system of values may govern someone's behavior without those values being fully explicit. They may figure in people's motivations and in their attempts to justify or criticize their own actions or those of others. Indeed, it may require a theorist or a therapist to make someone's values explicit. In this respect a system of values may be like an iceberg; most of what is important may be submerged and invisible, even to the person whose values they are. Because values are often opaque even to the person who holds them, there can be inconsistencies and incoherencies in a system of values. Indeed, much debate and dialogue about values involves attempts to resolve inconsistencies and incoherencies in one direction or another.

A system of values is generally a cultural construction rather than an individual one. It makes sense to talk about contemporary American values or those of eighteenth century England or tenth century India. Our individual differences on which we become highly polarized often tend to occur around the edges of our value system. The vast areas of agreement often are invisible because they are presupposed or assumed without argument.

I believe that our dominant value system is inadequate and inappropriate for guiding our thinking about such problems as climate change. Our dominant system, as it impinges on the environment, can be thought of as a relatively recent construction coincident with the rise of capitalism and modern science, and expressed in the writings of such philosophers as Francis Bacon and John Locke. It evolved in low population density, low technology societies, with seemingly unlimited access to land and other resources. This value system is reflected in attitudes towards population, consumption, technology, social justice, future generations, nonhuman entities and people across national boundaries.

The feature of this value system that I want to discuss is its conception of responsibility. Our current value system generally presupposes that harms and their causes are individual, that they can readily be identified, and that they are local in space and time. The English Common Law tradition theorizes many such cases. Consider the following, technologically updated, example.

Jones breaks into Smith's house and steals Smith's television set. Jones' intent is clear; she wants Smith's TV set. Smith suffers a clear harm. She's made worse off by having lost the TV set. Jones is responsible for Smith's loss, for she was the cause of the harm and no one else was involved. What we have here is a clear, self-contained story about Smith's loss. We know how to identify the harms and how to assign responsibility. We respond to this breach in our norms by punishing Jones in order to prevent her from doing it again, and to deter others from such acts; or we require compensation from Jones so that Smith is restored to her former position.

It is my contention that this paradigm collapses when we try to apply it to the climate change problem. It is for this reason, I think, that we often feel confused about how to conceptualize these issues.

There are three important dimensions along which the climate change problem varies from the paradigm. First, apparently innocent acts can have devastating consequences. Second, causes and the harms that result from them may be very diffuse. Third, causes and harms may be remote in both space and time. There are further features of the climate change problem that make it difficult for us to assess. These include various threshold effects, the nonlinear causation that is involved, and the irrelevancy of political boundaries to the distribution of physical effects.

Consider the following example. Some projections suggest that one effect of greenhouse warming may be to shift the Southern Hemisphere cyclone belt to the south. If this occurs, the frequency of cyclones in Sydney, Australia, will increase enormously, resulting in great death and destruction. The causes of this death and destruction will be diffuse. There is no one whom we will be able to identify as the cause of the destruction in the way in which we can identify Jones as the cause of Smith's loss. Instead of a single cause, millions of people will have made tiny, almost imperceptible causal contributions, by driving cars, cutting trees, using electricity and so on. They will have made these contributions in the course of their daily lives performing apparently innocent acts, without intending to bring about any harm. Indeed, most of these people will be geographically remote from Sydney, Australia. Many of these people will probably have no idea where Sydney, Australia, is.

Some people who are harmed will also be remote in time from those who have harmed them. Sydney may suffer in the twenty-first century in part because of people's behavior in the nineteenth and twentieth centuries. Many small people doing small things over a long period of time together may lead to unimaginable harms.

Despite the fact that serious, clearly identifiable harms will have occurred due to human agency, our dominant value system will have a very difficult time finding anyone to blame, for no one intended the bad outcome, nor brought it about, nor may even be able to foresee it. The fundamental moral crisis here is that we face the possibility of a serious disruption of the global environment, yet no one may be responsible.

This is a new problem. It takes a great many people and a high level of consumption and production to change the Earth's climate. It could not have been done in the low-density, low-technology societies in which our values evolved. London, for example, could be polluted by its inhabitants in the eighteenth century, and indeed was in a serious way, but its reach was limited. It is this global reach of our behavior that I think is new.

The natural question that arises here is this: if I am correct in suggesting that our traditional values are not adequate for addressing the climate change problem, then what sort of value system would be adequate? I do not have a full or compelling answer to this question. However, I will make some suggestions.

It seems to me that we would do well to return to some old-fashioned virtues such as those of conservatism and modesty. If each individual computes the effects of each action, each can correctly say, "It does not matter what I do, for whatever will happen with the global climate will happen regardless of my action." Although it may not matter what each individual does, it matters enormously what we all do together. In order to deal with this problem, we need to inculcate patterns of behavior that are noncalculating. We should try to cause people to do the right thing because it is right. Only then will we achieve the best outcomes.

Secondly, we need to think in a more holistic way. When pollutants come out of the tailpipe of our cars and are dispersed into the atmosphere, it is very easy for us to think that they cease to exist. But, in fact, pollutants go somewhere. Most pollution problems are problems of distribution. We need to appreciate the linkages in our behavior that exist across both space and time. Just as the emission of greenhouse gases makes us a community extended in space, so it also makes us a community extended in time. From our forebears we inherited an atmosphere loaded with greenhouse gases, and what we do now will affect future generations. Our thought needs to become as global in space and time as the effects of our actions.

As a society we have been moving towards a limited kind of global consciousness. Many of us are quite concerned about the policies that Brazil adopts with respect to Amazonia. We think of the Amazon as the "lungs of the Earth", and think of ourselves as having as much of a stake in its future as the Brazilians. But for most of us this concern is not brought together with domestic political decisions such as whether or not to open up the Arctic National Wildlife Refuge to oil exploration. Pumping the Arctic dry would permit us to continue business as usual for a relatively short period of time, but would have enormous long- term impacts on arctic ecosystems. It is important to see that Amazonia and the Arctic are linked; human actions in both areas are important to the future development of climate.

Finally, there is another aspect of our sensibility that we need to develop. I call this "mindfulness." In our society there are certain areas of human life that are highly moralized, for example sexual behavior. We often think of sexual behavior as being the paradigm of behavior that carries a moral charge or valence. Due to such problems as climate change, we are going to have to moralize or valence our behavior that impacts on the environment in much the same way that we moralize sexual behavior.

On the face of it this may seem impossible. How could such a transition ever be made? But consider an example. Two decades ago cigarette smoking was not a highly moralized or valenced behavior. Some people smoked, some people did not; it was a matter of personal preference. Some people may not have liked others to smoke in public, but such smoking was not thought of as imposing harms on others or as immoral. Yet nowadays this is what most of us have come to think, and this is what justifies the regulation of smoking in public places. It is possible to change the moral valence of an action in a very short period of time, as this example shows.

Most of what I have said applies generally to the sorts of values that we bring to the climate change problem. I will close with several comments specifically about corporate responsibility.

The first comment is that, although corporations have played a role in causing the climate change problem, they are not uniquely and decisively the bad guys. This follows more or less directly from the climate change problem being a highly fragmented, complex, collective action problem. Carbon is the lifeblood of industrial civilization as it has been practiced for the last two centuries. We all benefit from its use. While corporations have an important role to play in addressing the climate change problem, it is not a distinctive role relative to those of individuals, governments, and other forms of human organization.

Still, if corporations would adopt some of the reforms being urged upon them on other grounds, this would have significant environmental benefits. For example, if corporations would have longer time horizons and would internalize their externalized costs, this might help them to be both more profitable and more environmentally friendly.

Beyond this there are various other virtues which would contribute enormously to attempts to establish climate stabilization policies, if corporations would manifest them to a greater degree.

First, corporations should not be obstructionist when it comes to environmental regulation. Obstructionism is not a winning strategy for corporations in the long term, if for no other reason than that no one's long-term interests are served by obstructionism. In the short term such actions may be disastrous both for themselves and for society as a whole. Consider an example of obstructionism. The oil industry typically opposes "command and control" environmental regulations on the grounds that market approaches to addressing environmental problems are preferable. Yet the oil industry also opposes such market approaches as increased taxes on oil. They cannot have it both ways. If the oil industry is to be environmentally responsible, it must endorse some positive steps for dealing with the environmental impacts of the production and consumption of oil.

Second, corporations should be more flexible in responding to changing business environments. Japanese and German firms have seen that there are profits to be made in producing environmental control technology. Too often American corporations behave as if changes in the business environment are necessarily bad, and that the old ways of doing business can continue forever.

Finally American businesses have sometimes undertaken symbolic actions intended to demonstrate their concern for the environment. These actions should be applauded, but often are not. Symbolic actions are not usually taken very seriously; indeed they are often viewed as hypocritical. For example, it is sometimes said that corporations support the "green consumer" movement in order to avoid dealing with the real problems. But it can always be said about any particular action taken to address climate stabilization that it is unimportant or even hypocritical. Collective action problems by their very nature cannot be solved by a single dramatic action by a single powerful player. Any particular action will at best be a small step towards a solution. But solutions are built up out of such small steps, however inefficacious they may be taken by themselves. Symbolic action is important because it is often contagious (Kelman, 1990). Individuals and corporations who engage in symbolic action become role models for others. Symbolic action also puts a corporation on record as being on the side of environment-friendly change. Such actions send a message that a corporation is concerned about the problem, and wants to be part of the solution. By making such a statement a corporation makes itself accountable for its actions.

As I have tried to show, the climate change problem involves profound questions of personal and corporate responsible. Solutions will never be found in science and technology alone; they must also come from deep within the human heart.

REFERENCES

Committee on Earth Sciences, 1991, "Our Changing Planet: The FY 1991 U.S. Global Change Research Program," U.S. Geological Survey, Reston, VA.

Kelman, Steven, 1990, Why public ideas matter, *in:* "The Power of Public Ideas," Robert Reich, ed., Harvard University Press, Cambridge, MA: 30-53.

Lau, Richard, Brown, T., and Sears, David, 1978, Self-interest and civilians' attitudes toward the Vietnam War, *in: Public Opinion Quarterly* 42: 464-83.

Orren, Gary, 1990, Beyond self-interest, *in:* "The Power of Public Ideas," Robert Reich, ed., Harvard University Press, Cambridge, MA: 13-29.

Stern, Paul, Aronson, E., Darley, J., Hill, D., Hirst, E., Kempton, W., and Wilbanks, T., 1986, The effectiveness of incentives for residential energy consumption, *Evaluation Review* 10:147-76.

NOTE

[1]The next thirteen paragraphs are drawn from my "Ethics, Public Policy, and Global Warming," *Science, Technology, and Human Values*, 17, 2 (Spring 1992), copyright 1992 by Sage Publications, Inc., and reprinted with their permission.

INTERGENERATIONAL EQUITY

Douglas MacLean

Professor
Philosophy Department
University of Maryland
Baltimore County (UMBC)
Catonsville, MD 21228

I will not discuss how to allocate entitlements now and in the future but will instead point out the enormous complexity of this problem, involving both international and intergenerational issues. I'll consider myself fortunate if I can say something useful about the intergenerational issues. Bill Nitze talked about, and Dale Jamieson demonstrated, the need for a new way of thinking about ethics and future generations. I consider my remarks to be complementary to their comments and will discuss the same theme although I will focus on one particular aspect of it.

The problems are new. I think the approaches we need are also new and I wish to point out as well that philosophers have no ready-made and well understood expertise to help us with this new way of thinking, because these problems raise some new fundamental philosophical issues. We all need to think harder and more deeply about our moral values and about ethics. That is what I'll be demonstrating here, I hope.

Any discussion of the topic of intergenerational ethics or intergenerational justice in the area of environmental ethics and climate change must, it seems to me, begin with the fact of uncertainty. One of the major questions is whether it has to end with that fact as well. I think it does not, but I will be travelling down a rather circuitous road in discussing uncertainty to show where I think there is hope for solutions, and the road may take some surprising turns.

It might help to begin with two different kinds of uncertainty, physical uncertainty and what we might call historical uncertainty relating to issues of climate change. I am sure most of you understand physical uncertainty at least as well as, and many of you better than, I do. I will give several important examples, one of which has to do with feedback. All of the intergovernmental models on climate change which give projections for greenhouse gases to the year 2100 show carbon dioxide, nitrous oxide and CFCs increasing. But all of these models ignore a large range of possible feedback effects on the concentrations of gases which result from

Global Climate Change, Edited by J.C. White
Plenum Press, New York, 1992

global warming itself. They are likely to be positive, that is, to increase greenhouse gas concentrations, but we do not understand how they operate. This is one major cause of uncertainty in the models.

Another is that the effects of greenhouse gases on climates have been predicted only by general circulation models. These work with grid points spread across the globe and equations on the models are solved at these points only. Because the models make huge demands on computer power, the grid cannot be very fine and some important atmospheric phenomena, such as cloud formations, take place on a smaller scale. Cloud cover is a vital ingredient in predictions of global warming because clouds may cool the atmosphere by blocking out radiation from the sun, or they may warm the atmosphere by trapping radiation from the earth's surface. The methods of estimating these effects, called parameterization, is controversial at its core, and it can lead to large uncertainties in the model. The examples of physical uncertainties at this level can, of course, be multiplied.

In addition, there are what I want to call historical uncertainties. Even more profound than physical uncertainties are the uncertainties involved in translating the estimates of climate change into estimates of effects on society. Thomas Schelling, in an essay written more than a decade ago, asked us to imagine trying to make predictions for today as if we were in 1900, which is not too dissimilar from what we may be trying to do looking today at the year 2100 and beyond. He reminds us that electronics were not dreamed of, transatlantic travel by zeppelin was to be possible within a generation in the future, Russia was czarist, U.S. life expectancy at birth was 47 years, and only one third of the U.S. population lived in places with more than 500 inhabitants. The task of prediction is formidable.

In addition, there are several more reasons why historical uncertainty creates special problems. One is that climate change itself helps to determine history. Consider, for one example, how the little ice age in medieval times caused the Vikings to retreat from Greenland. But for this event, America might have been settled by Danes from the North rather than by Spaniards from the South and, of course, it could have been quite a different place.

Another reason is that technological developments occur in unpredictable ways. Technology today has made us perhaps less dependent on climate than we were ninety years ago, but technological fixes are certainly limited, and some influences on climate and on history would appear to be unaffected by technology.

Schelling asks us to suppose that we had perfect facts of climate change to 2100. There would be places in important parts of the world where we could not even put an algebraic sign on the apparent welfare aspect of climate change, let alone know its magnitude. As he writes, "Unless we impute to ourselves foresight much superior to what we might willingly claim for ourselves were we doing our work in 1900, it is likely that most of the identifiable changes in welfare due to climate change would be, for most parts of the world, swamped by other uncertainties." Nevertheless, many other writers, like John Broome in England, argue that we can still fairly safely predict that the effects of global warming will be long-lived, almost certainly large, probably bad and possibly disastrous.

A third aspect of historical uncertainty is the effect of climate on population, both on demography and on population size. Changed social patterns will affect birth rates, though we may not know how. Part of this change also involves what philosophers following Derek Parfit would call "changing the identity of future generations." By this, we mean that changing social patterns affect not only the numbers of people who will be born in the future, but also affect the timing of those births. When something changes the timing of births in unpredictable ways, it determines that one population will exist in the future rather than another, for when you change the timing of a birth, you change who will be born.

Now this, although it seems odd, is an indisputable effect of massive changes like climate change. I think it's also very important in understanding our intergenerational moral responsibilities. I will try to explain how. Climate change will affect not only the number of future generations, the number of people within a future generation and the environment of future generations, but it will also affect the identities of people within future generations.

I will call this population effect the identity problem. It is important, not as some have suggested because it shows that we will not have any moral obligations to future generations, but because whatever we do will determine who will exist then. I think that this problem tells us something about how we can and how we cannot understand the nature of our moral responsibility, the foundation of intergenerational ethics. It is important, first of all, to understand these things correctly, to have a coherent view of morality. Also I think a coherent view of morality in this area might serve us well in leading us to some more practical conclusions.

So let me explore that very fundamental foundational issue for a minute. What has posterity done for me? Why do we owe anything to the future at all? If this question means, "what is my debt to posterity?" then the answer of course is nothing. But this is a trivial answer to a trivial question, for we can't have a debt to something in the future. The notion of debt in this sense is something that is connected to the direction of time, so the fact that we owe future generations nothing merely because they exist in the future tells us nothing more than that we are located earlier than they are, which is not morally important or important for any other reason.

If the question, "what has posterity done for me?" means "is it to my advantage to care about future generations?" then I think we have to look a bit more seriously at it, although it's still not well formulated. Will we benefit from caring for future generations? Well, we may or we may not, but why should this be the fundamental moral question we ask? After all, we do not expect morality in every case to advance our immediate self-interest or to work to our advantage in our normal lives. Why should we expect it to here?

Is it in our interest to care about future generations? I think it is, and we need to explain why. But before we consider the nature of our interests and why we are and ought to be interested in future generations, let us look again at our moral obligations.

Most moral theories that one would think of applying to these questions are, as Dale Jamieson has explained quite clearly, theories that were devised for other purposes at other times, under other circumstances. Theories based on rights and justice, including most contemporary social contract theories, rely on what we can call person-affecting principles. That is, they are about ways of affecting other people. They assume that there is a person, discuss different ways we might or can treat that person, and ask whether some ways are acceptable or unacceptable. This is a normal way to think about intergenerational responsibility.

Let me give an example from one well-known philosopher whose views are unassailable on most moral issues, and whose view on this issue reflects quite a bit of common thinking. Joel Feinberg writes:

> Every paper must begin with an unproved assumption. Mine is that there will still be a world 500 years from now, and that it will contain human beings who are very much like us. We have it within our power

now clearly to affect the lives of those creatures for better or worse, by contributing to the conservation or corruption of the environment in which they must live. From the perspective of our remote descendants, it is basically a matter of justice or respect for their rights. But if the identity problem is true, one thing we clearly do not have is the power to affect the lives of these identifiable individuals for better or worse.

The power we have is to create some individuals living under good circumstances or other individuals living under circumstances we would consider unfavorable. If there is one thing any acceptable theory of rights or justice does not tell us, it is that advantaged people have a right or a demand of justice that they, rather than disadvantaged people, exist. I think that this normal, very natural way of thinking cannot even begin to explain why we are or ought to be concerned with future generations.

Now, traditional utilitarianism can sidestep this problem, but there remain other ones. I will not go into them now. Let's follow another suggestion just for a minute, and look instead at the past, following Edmund Burke and a more traditional social contract theory. There's an intuitively appealing idea that we owe to future generations not what we owe to them, but what we owe to our predecessors, who after all worked, sacrificed, built our society and institutions, our economy and infrastructure—for what reasons? Not to pass it onto us so that we could deplete it all, but to create something that would be lasting. This is the idea that we are and ought to be stewards carrying on what they gave to us, improving it, and passing it on.

But the mere fact that they sacrificed for us cannot oblige us to their values. After all, some of the values of our ancestors are not values we are proud of or want to pass on. This fact, although it gets to an important insight, is also insufficient to explain the nature of our moral obligations. We have to examine the values directly. So let's look then at our own values.

There are two ways of understanding why it is in our interest to have a concern about future generations. One is that we have a preference for future generations, and in a democracy we ought to act on that preference. We have intertemporal preferences, and these must be respected. The problem is that these preferences do not take future generations seriously enough.

However, our values are not directed only at our own experiences. It's important to realize—and this is the first thing that we have to realize—that our values are also outward-directed. We care not just about our own happiness, but we find and increase our happiness by caring about things outside ourselves. This is a very important aspect of morality. We care about culture, we care about science, we care about increasing the store of knowledge, about building a better society, and we take our happiness in the thought that we participate in the development of this kind of world and improving it.

It's also necessary for us to have this outward-directedness in our values in order for our lives and our own projects to seem meaningful. If we imagined ourselves creating a world which gave us everything we wanted but which would cease to exist immediately upon our death, then I think that no world we could imagine creating would be one that didn't seem empty. To give us a sense of having some purpose, there have to be values that we think of as continuing into the future.

So the fundamental starting point, I would argue, is that the place to look is not at welfare or happiness, but at the more basic questions: What makes life meaningful? What kinds of values do we have? Here I think that we can begin to answer this question by seeing ourselves committed to certain aspects of culture, civilization and society, that we attempt to create not just for us, but because we consider these things in and of themselves good and worth preserving. This would tell us then that what we need to pass on are not only these institutions, but also the opportunities and resources for future generations to carry them on.

I think that this will have some fairly specific applications as to how we think about these issues. I can't go into them now, but it will tell us, for instance, what we would be reasonable to discount and not to discount in terms of welfare, benefits or values as they extend through time. They will also tell us something important about institutional design. We must offer ways of protecting and ways of debating, discussing and formulating these important outward-directed values, and then designing policies to reflect them. The idea here is that our issues are not all economic issues, but political as well.

The debates we will face all have ideological and political components that we must recognize. Their importance goes far beyond wondering how we are to define the technological fix that will fuel our further economic growth without making the planet uninhabitable. We cannot resolve these debates with refined economic or analytic techniques alone because they cannot address the underlying question which matters to us most: What kind of world are we trying to create?

DISCUSSION

THE ETHICS OF LIVING IN A GLOBAL GREENHOUSE

NITZE: This is my response to what Doug MacLean just said. Actually, his definition of happiness has very ancient roots. It goes back to the writings of Aristotle thousands of years ago. I think that's a very convincing place to start, because if we accept the Aristotelian definition of happiness, much flows from it. But where we have to start is with knowledge. I think our knowledge is still insufficient in many cases to be as morally sophisticated as we could be about achieving happiness. There really is an interplay here between knowledge and ethical behavior, and hopefully increased knowledge will lead to real changes in behavior.

QUESTION: You suggest a need for different values, a values change without parallel change in the economic systems, political systems, technology, et cetera. If such parallel and simultaneous changes are needed to buttress the new values, what must be the nature of those additional changes?

JAMIESON: It is shocking how little we know about value change. Part of the reason for that is because this whole area of value change has not been much studied empirically or even conceptually. My response to this question would be to say, yes, I agree. I think there are probably very complicated interactions between the way moral systems, economic systems and political systems change. I think there's a tendency for people to think the real action is economic, and then the barroom philosophers come along afterwards and rationalize the changes.

But I think it's actually more complicated than that. I would again point to the example of cigarette smoking. I think, in that case, the motor behind changing laws and regulations was a more fundamental moral change that people had about cigarette smoking.

QUESTION: The writings of Joseph Campbell forcefully define the fundamental differences between the underlying value systems of the East and West. How can we create a global environment oriented value system which bridges and encompasses these differences?

JAMIESON: First, one of the consequences of trying to be interdisciplinary is that one can be almost equally ignorant of all fields. I confess to not having read Joseph Campbell. Nevertheless, the need to combine different cultures' value systems is an important problem.

I would give two remarks on this. One is that, just as the moral disagreements that we have within our own culture tend to blind us to the broad amount of agreement which remains invisible because it's taken for granted, likewise I think that, as different as the value systems of different cultures are, they are bound to be based on a very wide area of agreement.

Having said that, I think there are still important differences. I think it's increasingly imperative for us and for the philosophy profession in the English-speaking world to pay more attention to the ethics and philosophies of other cultures which are not tied so closely to the development of capitalism and the kinds of economic growth and situation that ours has been tied to. We especially have a lot to gain by taking these other value systems seriously. That's a responsibility that I think can be discharged best beginning with academics and philosophers in the courses that we teach.

QUESTION: How would you define the role of the federal government, especially the President, in shaping peoples' values on environmental issues?

JAMIESON: I guess I think that in this whole area there are lots of players. The President is one player but not the only player, nor is the President even quite as decisive on this issue as he is on many other issues. Also presidents can play a lot of different roles. One role that presidents can play is a leadership role. That is, the president can encourage the development of new values and different perspectives on the part of a population. But a president can also perform another symbolic role—he can be the whipping boy. He can be the opposition point, so to speak, such that peoples' positions and values are oriented towards critique and criticism of the president's position. Obviously presidents are important, and so are a lot of other people on this issue.

QUESTION: How do we know when we have saved or invested enough for future generations? Is it meaningful to talk about discharging our obligations where future generations are concerned?

JAMIESON: Of course it's difficult to know. But I think it is meaningful to talk about discharging our obligations. It's going to be very hard to answer your question if we think of our obligations as having to do with welfare in the future. We should think instead about what our obligations might be to our own immediate children. I don't think that I have an obligation as a father to my children to ensure their happiness. I think my obligation instead is to try to create in them a reasonable value system, and then to pass on to them the opportunities to act on that. Beyond that, if one of my children wastes it all away and the other makes good with it, that's really out of my hands and beyond my responsibility.

So I take seriously the idea that we want to create a culture that will give values that future generations will want and see as reasonable to embrace, and then to provide them not with welfare, because we can't control that, but with sufficient opportunity and resources to act on those values and to build on that culture.

I would like to make two comments. First, I think older cultures from different parts of the world have much to teach us, including our own Native American culture, about how to live in a world where humanity is a little bit more humble about its degree of control over natural systems that surround it. We have a great deal to learn, not only from Native American culture, but from cultures in other parts of the world which do take a more synergistic and cooperative view of human societies and their relationships to natural systems.

Secondly, there is within our own culture a history of a very powerful myth which I think will help us to find our way out, and this is the myth of the fall. The myth of the fall is, in the Judeo-Christian tradition, the fall of Satan himself; this story is told in the Book of Genesis. We have modern counterparts to this great story in science fiction. All of you are familiar with Star Wars and George Lucas and the creation of the figure of Darth Vader, who is indeed one of the Jedi knights, only a fallen one. So the history of the fall is a bit of a corrective mechanism within our own culture, and it runs very deep. The Wagnerian legends point to another aspect of it, which is that the fall is associated with the desire to assert dominance, to be God. Science and technology are moving into a position where we are asserting true dominance and playing a truly godlike role. I think there are certain checks and balances within our own culture which perhaps will bring us back to some sanity on the degree to which and how we play God.

Examining the Linkages—Which Policies Will Work?

ENERGY POLICY: INSURANCE AGAINST THE UNKNOWN

David Bodansky

Professor
Department of Physics
Department FM-15
University of Washington
Seattle, WA 98195

In considering appropriate responses to the problem of global climate change, I would like to suggest three key points:

(1) Prudence calls for making vigorous efforts to reduce greenhouse gas emissions, with the initial burden perforce falling on the industrialized countries. This point is somewhat controversial.

(2) The reduction of the use of fossil fuels is central to addressing the problem of greenhouse gas emissions. This is not very controversial.

(3) In implementing future energy policies, it is a mistake to neglect the expansion of nuclear power, along with conservation and solar power. The nuclear issue is so controversial that it sometimes is ignored as being too controversial to address.

The establishment of suitable climate change policies is complicated by many gaps in our knowledge. The greenhouse gases have complex sets of sources and sinks which are not well understood. There are great uncertainties in predicting the climate changes which may be caused by changing atmospheric concentrations of greenhouse gases. It is also very difficult to anticipate the abilities of societies and the biological environment to adjust to whatever climate changes do occur.

Matters are further complicated by the fact that, if we wait for an incontrovertible signal of a temperature rise, we will have waited until very late in the game. First, there is a time lag of perhaps several decades between the emission of greenhouse gases and the resulting temperature rise. This arises from the thermal inertia of the oceans. Second, if a compelling signal is seen it would be impossible to reverse quickly the direction of an energy economy built on fossil fuels. Implementing large-scale changes must be a gradual and incremental process. If needed, it is best started before the need becomes urgent.

Global Climate Change, Edited by J.C. White
Plenum Press, New York, 1992

Given the uncertainties, scenarios for the future vary from the relatively benign to the disastrous. Suggested responses range from calls for immediate strong action, on the grounds that it is better to be safe than sorry, to suggestions that we wait and see, studying the problem before taking precipitous, expensive actions. Intermediate between these are the "no regrets" policies, in which actions are limited to those which have merit on their own, particularly energy conservation.

There are considerable differences between this spectrum of responses and the way society responds to other threats. For example, under the Delaney Clause a food additive is banned if there a hint that it may cause cancer. One does not wait for proof; very incomplete evidence suffices. Some argue that this is an example of over-reaction. But always the problem in matters of risk lies in finding the proper balance between excessive reaction and too little reaction. It is the premise of these remarks that, in the case of global climate change, plausible consequences could be so severe that the prudent balance lies on the side of vigorous action. This is a better-safe-than-sorry policy, rather than a no-regrets policy. The difficulty with a no-regrets policy is that only in retrospect does one know what might turn out to be a cause for regret. The concept of insurance is relevant. Whether a particular insurance policy was or was not worthwhile is only established after the event. But there is no actuarial experience or other firm evidence on which to base a decision as to the size of a premium one should pay for the insurance against anthropogenic global climate change.

There is a model from far afield for how the greenhouse gas emissions might be approached. The internationally adopted policy for protection against ionizing radiation is to keep exposures "...as low as reasonably achievable, economic and social factors being taken into account" (1). This prescription in one sense is very vague and imprecise, almost uselessly imprecise. But nonetheless, it sets a clear sense of direction. It establishes a standard under which it is necessary to justify radiation exposures and reduce them when possible. A similar policy for greenhouse gas emissions would strongly influence decisions on the acceptability of differing energy policies, weighting the scale heavily in favor of alternatives to fossil fuels.

A major problem in addressing global climate change arises from the complexity of the international issues of equity. Countries differ greatly in the amounts they contribute to greenhouse gas emissions, in the rates at which their emissions are changing, in the possible effects of global climate change upon them (the winners and losers question), and in the resources they have to reduce emissions or cope with a changed future environment.

We here focus on carbon dioxide, which deservedly receives most of the attention. Roughly one-quarter of the world's anthropogenic CO_2 is being emitted by the United States, one-quarter by the other members of the Organization for Economic Co-operation and Development (OECD) (essentially the industrialized countries outside the former Soviet bloc), one-quarter by Eastern Europe and the former USSR, and one-quarter by the developing countries. The developing world, which contributes about one-quarter of the CO_2, has about three-quarters of the population. There is an enormous discrepancy in per capita emissions between the developing countries and the United States, for example a discrepancy in 1989 of a factor of nine between China and the United States and of about a factor of twenty-five between India and the United States (2).

At the same time, the rate of increase of fossil fuel use is much greater for the developing countries than for the industrialized ones. There is a strong correlation between the growth in energy consumption and the energy consumption per capita. At the extremes among populous countries, the 1989 per capita consumption of commercial energy in Bangladesh was under 1% that of the United States, but it increased during the 1980s at almost eight times the rate (3). The combination of population growth and increasing per capita energy consumption in the developing world may make the problem appear intractable.

But we do not know what will prove to be intractable several decades hence. The question is what should be done now. With the industrialized world, including the countries of the former Soviet bloc, responsible for three-quarters of the production of CO_2, the steps they take can have a very large impact over the next twenty or thirty years. Not only are these countries the main producers of CO_2, they also have the greater resources and flexibility to make changes. Without addressing all the issues of equity, it is appropriate for them to assume the initial burden. Their actions alone will not solve the problem for the next century, but can greatly influence the progress the world makes over the next several decades.

In considering what constitutes effective energy policies, it is useful to look at the recent record. Figure 1 shows changes in the period from 1973 to 1989 in energy consumption, oil consumption, and CO_2 emissions for the world, for the OECD countries, for the United States, and for France. For the world as a whole, CO_2 emissions rose 28% in this period, while for France they dropped 28%, despite a substantial increase in total energy use. Carbon dioxide emissions rose 4% in the United States and were about constant for the remainder of the OECD. For the world outside the OECD, there was a 57% increase in CO_2, reflecting the increased energy consumption in the less industrialized countries.

It is puzzling to see references to a goal of stabilizing emissions at the 1990 levels when one has the example of France and other countries which have already achieved marked reductions. The record of France, where CO_2 emissions decreased at a rate of better than two percent per year, is attributable largely to its intensive

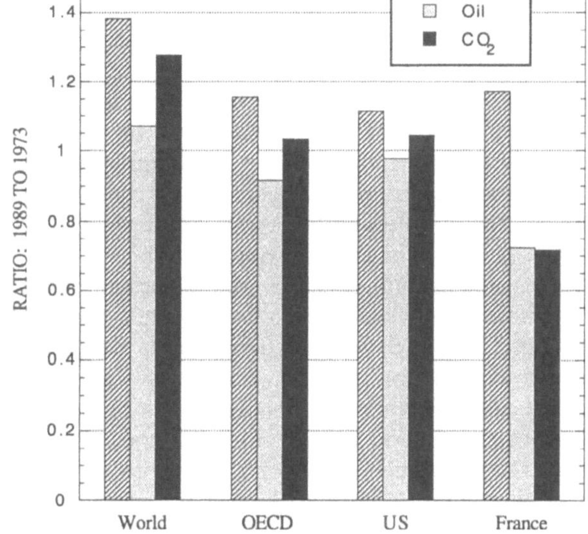

Figure 1. Ratio of energy consumption, oil consumption and CO_2 production in 1989 to values in 1973, for selected regions and countries [copied from Ref. (8)].

use of nuclear power. Nuclear power now provides France with about 75% of its electricity and more than one-third of its primary energy. There is nothing unique about the French situation, other than France's willingness to use nuclear power. Any industrialized country that wanted to emulate France could have. In fact, some followed a similar course. Sweden, Switzerland and Belgium also obtained large fractions of their electricity from nuclear power and achieved substantial reductions in CO_2 production.

Increased use of nuclear power is only one part of an effective approach to the reduction of fossil fuel consumption. Conservation is clearly important. It has had significant successes and it would be a tragedy not to implement its further potential. However, no matter what successes are achieved by conservation, new energy sources will be needed to replace fossil fuels and, in some cases, to meet new demands. To rely on conservation alone, and still generate electricity or heat buildings with fossil fuels, is to set an unnecessary limit on the proper ambitions of a greenhouse-gas energy policy.

The main alternatives to fossil fuels are solar power in its many forms and nuclear power, perhaps at some future time including fusion. Solar power should be pursued with increased efforts. However, despite its apparently large potential, it has had a disappointing record in the industrialized world since 1973, as can be seen from data on electricity production in the OECD countries (4). Suitable sites for additional hydroelectric power are rare in the OECD countries and hydroelectric's share of OECD electricity generation dropped from 21% in 1973 to 17% in 1989. In the same period nuclear's share rose from 5% to 23%. Non-hydro renewable resources, including geothermal, wind and other forms of solar power, contributed less than 0.3% of OECD electricity generation in 1989—mostly geothermal. There may be substantial prospects in wind and photovoltaic power, and perhaps in biomass in countries with available land, but these are still unproven hopes. There is no renewable energy analog to the record of nuclear power in France as an example of a large-scale success story.

This is not an argument for neglecting solar power because, as is often pointed out, part of the lag in solar power is due to lack of investment. The slow progress of solar power to date can be taken as an argument for more investment in it. But it is also a cautionary warning against assuming that dramatic future successes are guaranteed.

We now turn to nuclear power. Nuclear power was used in 1991 in 25 countries, with a total of somewhat over 400 reactors. In 1990, it supplied 17% of the electricity worldwide and more than 40% in six countries, led by France at about 75%. However, the further expansion of nuclear power is now almost halted, except in France (which is nearly saturated), Japan, South Korea, and, much more speculatively, in some countries of Eastern Europe.

There are three fundamental concerns about nuclear power. One is reactor safety. Chernobyl demonstrated vividly that a large nuclear accident can have major, worldwide effects. But Chernobyl was exceptional as a dangerously designed and dangerously operated reactor. In the end, one's view of nuclear reactor safety depends in large part upon whether one believes that Chernobyl is prototypical or was unique to the Soviet situation at the time. For the "Western" countries, the overall reactor safety record has been very good, even considering Three Mile Island. Future reactors can be made safer still, if there is the motivation to build them.

A second concern involves waste disposal. There is no permanent repository anywhere in the world for high-level wastes from commercial reactors and it is unlikely that any will be in operation in less than twenty years. Nonetheless, it is widely believed that the problems are more institutional than technical. As expressed in a recent statement from a committee of the National Research Council (5):

> There is no scientific or technical reason to think that a satisfactory geological repository cannot be built. Nonetheless, the U.S. program, as conceived and implemented over the past decade, is unlikely to succeed.

Complicating the process of finding a satisfactory solution is the opposition of those who, in the words of Michael McCloskey, Chairman of the Sierra Club (6), want to use the nuclear waste issue to "...drive a final stake in the heart of the nuclear power industry before they will feel comfortable in cooperating fully in a common effort at solving the waste problem..." If this is an accurate assessment, it describes a regrettable situation. What is needed is a cooperative effort to solve the problem.

Finally, there is the issue of weapons proliferation. In principle, weapons development can be furthered by wider availability of the expertise and equipment which go with any sort of nuclear reactor program. However, commercial nuclear power has not been a significant historical precursor to weapons development. Every country that admits to nuclear weapons had the weapons well before they had civilian nuclear power and some of the other countries most talked about in terms of having weapons or weapons ambitions have no civilian nuclear power at all, including Iraq, Israel and North Korea. In the world as it is, it is difficult to see how the abandonment of nuclear power by the United States or Europe would appreciably lessen the chance that other countries will develop weapons. The positive coupling between civilian power and nuclear weapons is weak and there are couplings in the opposite direction as well. With energy problems, there can be dangerous tensions. Witness the war with Iraq, which in part arose from the desire to gain or protect oil supplies.

In the end, we are left with competing fears as to the risks to ourselves and to future generations. However, if climate change constitutes a truly serious problem, we cannot afford the luxury of writing off nuclear power, the energy source which so far has had the greatest impact in reducing CO_2 emissions. It is intellectually frivolous to try to meet so great a global environmental threat without a careful evaluation of all mitigation measures, including an assessment of the costs, dangers and benefits of nuclear power. I suggest that such an evaluation will confirm that nuclear power is needed, along with conservation and solar power. These components of a rounded energy policy may play different roles in different countries, but all should be pursued.[1]

NOTE

[1]Many of the topics discussed in this paper are treated in more detail in References (7) and (8).

REFERENCES

1. ICRP, Recommendations of the International Commission on Radiological Protection, ICRP Publication 26, *Annals of the ICRP* 1, No. 3 (1977).
2. Gregg Marland and T.A. Boden, CO_2 Emissions—modern record/national, *in:* "Trends '91: A Compendium of Data on Global Change," T.A. Boden, R.J. Sepanski, and F.W. Stoss, eds., ORNL/CDIAC-46, Oak Ridge National Laboratory, Oak Ridge, TN (1991).
3. World Bank. "World Development Report 1991," Oxford University Press, Oxford (1991).
4. OECD/IEA. "Energy Balances of OECD Countries, 1980-1989," OECD/IEA, Paris (1991).
5. National Research Council. "Rethinking High-Level Radioactive Waste Disposal: A Position Statement of the Board on Radioactive Waste Management," National Academy Press, Washington DC (1990), p. vii.
6. Luther J. Carter. "Nuclear Imperatives and the Public Trust: Dealing with Radioactive Waste," Resources for the Future, Washington DC (1987), p. 431.
7. David Bodansky, Global Warming and Clean Electricity, *Plasma Physics and Controlled Fusion* 33:1489-1507 (1991); Proceedings of the 18th European Physical Society Plasma Physics Division Conference (Berlin, June 1991), D.C. Robinson, ed.
8. David Bodansky, Implementing the Nuclear Option, *in:* "Proceedings of the Conference on Global Climate Change and Energy Policy" (Los Alamos National Laboratory, October 1991), Louis Rosen and Robert Glasser, eds., American Institute of Physics, New York, NY (to be published, 1992).

DESIGNING A CARBON TAX STRATEGY

Roger Dower

Director
Climate, Energy and Pollution Program
World Resources Institute
1709 New York Avenue, NW
Washington, DC 20006

As one who thought he knew a lot about climate change issues, I find that there is always much more to learn. I have to say, as an economist, that I have been particularly encouraged by what I see and hear as a growing willingness to look at the role of the price mechanism as part of a response to the risks of climate change and as part of a move toward what I think is a broader energy transition. I view energy price reform, through carbon taxes in the case of climate change, and removing price subsidies as the core of any real effort to address climate change risk. That is not to say they are the only elements, by far. We have been reminded of the important role of values in that system, for example, but we also have to remember the role of prices in directing our energy consumption and production decisions and their risks.

Notwithstanding their economic appeal, pricing policies in general have never been a very popular way of dealing with environmental problems. Pricing policies, either through taxes or removing subsidies, have never in the U.S. or anywhere else in the world been a preferred route in terms of addressing environmental risks. Yet, we recognize all of the very important contributions that pricing can bring to the environmental debate, particularly in terms of the climate change issue. Again, we are reminded of the many hundreds and thousands of individual decisions that relate to the risks of climate change. We need to be looking for a centralized policy control mechanism. Prices, as such a mechanism, are often left out of the picture, in part because of perceptions of burdens and equity and notions of fairness that are often associated with price changes.

This often translates into a willingness to admit that prices can play an important role, but an unwillingness to use them because prices seem too burdensome; we need to look at other alternatives. I would like to argue that an analytical base is being developed to turn that argument around, to say that, with

Global Climate Change, Edited by J.C. White
Plenum Press, New York, 1992

careful design and construction, we can both get the benefits from good pricing policies and deal with the issues of burdens and equity.

What I would like to do then is to tie together some threads that have been introduced in earlier papers, maybe add a couple of others, and to try to construct a case for a pricing strategy, rather than just prices, as a way of beginning to deal with the risks of climate change. I will do this by using carbon taxes as an example with a specific U.S. focus. But a number of these principles are applicable to a much wider range of countries, both industrialized and developing, and to a wider range of policies.

Let me start with what I think is a principal myth of energy price reform efforts. The myth is that reforming energy prices always hurts—it hurts people, it hurts industry, it hurts the economy.

First, look at what subsidized or artificially low energy prices have brought to Eastern Europe and many other parts of the world which are highly reliant on coal. Air, water and land pollution in these countries is simply staggering. One sees quite directly that reforming energy prices, in terms of letting energy prices better reflect the social costs of that consumption, does not necessarily mean you have to hurt. There is a problem, however, when we do not measure the impact of the costs of pollution within our traditional accounting framework. We end up measuring pain, when in fact price reform can actually create real gain.

Having stated the broad macro picture, it is equally important to note that energy price reform does create economic change and transition, thereby creating winners and losers. There is no way out of it. These changes do not need to be permanent, but they do occur. Of course, such changes are not unique to energy taxes. They occur with any serious effort to reduce carbon emissions. Emissions trading programs and standard regulatory programs create economic change and transition, create winners and losers. They are not as explicit—they are more hidden—but we need to remember they are there. Using price reform as a basis for reducing pollution, however, offers an advantage over the alternative approaches in the sense that they create revenues. These various concerns over energy price reform create the basis of thinking about a carbon strategy. They suggest some key elements that we need to keep in mind if we are going to take advantage of what energy prices can do in an energy transition, and as well deal with the burdens associated with it. I will discuss three parts of this energy price reform strategy that I think are important.

First, if we know we're going to create economic change and create economic risks, let's undertake efforts to manage the economic risks. Lots of opportunities exist, but let me suggest three. First, you can adjust other energy taxes. That is the European experience, in large part. It is commonly held that Europeans have much higher energy prices than we do in the U.S. That's true to an extent. They have higher gasoline prices but not necessarily higher energy prices across the board. Because Europeans are as concerned about energy prices as we are, they have tended to adopt new energy taxes only by trading one set of taxes for another.

By way of example, Table 1 shows the Swedish carbon tax proposal. They have created a new energy tax, added on a CO_2 tax, tacked on a sulfur tax, and reduced other taxes. As a result, they ended up with a net no-change in the gasoline tax, but with an increase in several of the other taxes. They are adjusting the mix of energy taxes as a way of reducing the economic impacts of a CO_2 tax, while simultaneously increasing the environmental benefits.

The opportunities for industrialized countries to trade energy taxes as a way of adapting to the carbon taxes are significant. Table 2, prepared by the OECD,

Table 1. Fuel Price Implications of the Swedish Tax Reform (US$)

	Excise Taxes Before and After Tax Reform				
	Energy Tax	CO_2 Tax	Sulphur Tax	New Tax	Change against December 1990
	In January 1991				
Gasoline (1,000/litre)	488.80	91.80	••	547.40	0
Diesel Oil (1,000/litre)	91.80	122.40	••	214.20	+54.40
Light Fuel Oil for Industry (1,000/litre)	91.80	122.40	9.18	223.38	+40.12
Light Fuel Oil for Households (1,000/litre)	91.80	122.40	9.18	223.38	••
Heavy Fuel Oil for Industry (metric tons)	91.80	122.40	36.72	240.72	+67.66
Natural Gas for Households (1,000m³)	29.75	90.95	••	120.70	+61.20
Steam Coal Price, Industry (metric ton)	39.10	105.40	38.25	182.75	+104.55

Table 2. Implicit Carbon Taxes in 1988 ($ per ton of carbon)

	Oil and Oil Products	Gas	Coal	Total
United States	65	0	0	28
Japan	130	2	0	79
Germany	212	23	0	95
France	351	38	0	229
Italy	317	80	0	223
United Kingdom	297	0	0	106
Canada	108	0	0	52
Austria	267	39	0	150
Belgium	162	35	0	86
Denmark	297	110	0	147
Finland	189	0	0	107
Ireland	277	4	0	138
Netherlands	221	27	0	89
New Zealand	235	0	0	117
Norway	258	0	0	182
Portugal	205	131	0	150
Spain	176	19	0	112
Sweden	268	13	6	214
Switzerland	224	2	18	198

translates existing energy taxes in Europe into carbon tax potentials. That is, if existing energy taxes were based on carbon, this is what they would look like. With the exception of the new carbon tax in Sweden, existing taxes are all on gasoline, some natural gas, and virtually none on coal. The existing taxes on gasoline are equivalent to carbon taxes of hundreds of dollars per ton of carbon. No one has been talking about this level of carbon taxes, at least in the short run. So the possibility of shifting existing tax burdens from gas to coal and oil is one way of managing economic risks.

Second, you can adjust the basis and the size of the tax. That's another approach that is used commonly in Europe. The current version of the EC energy/carbon tax does not apply the tax on certain industries that are particularly sensitive to energy price increases. Of course, all of that affects how much you get in the way of CO_2 reductions and at what cost.

Finally, the tax can be phased-in. Energy tax reform proposals in this country are often haphazard—let's just go in and place a new tax on the economy. It is little wonder that, when you do that, you find everybody's models of the macro-economy showing huge economic losses. But phasing in a tax makes a big difference and can help lower the economic costs.

Let me move to a second element of any carbon tax strategy, specifically, the economic benefits of recycling the revenues. Other papers have presented some of this data and I will not go into many of the analytical details, but I hold it out as a separate item, because it's an attribute unique to the price reform approach to coping with the energy transition. The general point is that a carbon tax creates revenues that can be put to good use. You can recycle them into the economy and make a difference. The EPA paper, in this volume, shows us how recycling the revenues can make a difference in terms of the economic risks. Lowering other taxes as you raise carbon taxes will lower the impacts on GNP and lower the effects of the carbon tax on inflation.

I will show you a different set of numbers in Table 3 from an analysis done by the Department of Energy, using the Data Resource, Inc., model and demonstrate the difference it can make between not recycling and recycling the revenues. These estimates are in terms of reducing payroll taxes. There is a moderating impact on the economy no matter what type of energy tax you start with. In terms of the carbon tax, the impact on GNP over a ten-year period can be reduced by almost ninety percent. You can play with the models and get different kinds of results. It is the concept that is important.

Table 3. Estimated Impact on GNP of Various Types of Energy Taxes

Impacts of Real GNP		
	Deficit Reduction	Deficit Neutral
15% Ad Valorem Tax	-277	-51
$1.00 Btu Tax	-266	-41
$1.00 Carbon Tax	-226	-23
50¢ Motor Fuels Tax	-218	-59
Phased-in Motor Fuels Tax	-161	-37

(NPV, 1991-2000, 10%)

More importantly, however, is the opportunity for reducing the economic cost of taxation by raising carbon taxes and lowering other taxes. The way we raise revenues in this country costs a lot. I will only show a quick set of estimates. Figure 1 shows that every dollar that we raise by taxing capital, for example, actually costs the economy 46 cents. There are other estimates like this in the literature and some are much higher. The point is that we can shift the tax burden in our economy—the tax shift that we talked about before—and gain both in terms of reducing the impact of our existing tax system and improving the environment. That is part of what was called a no-regrets tax policy in an earlier paper. We can shift existing burdens and improve our economic and environmental position.

Figure 1. Efficiency costs of U.S. taxes.

The final element of a carbon tax strategy goes back to an earlier point that pricing reform creates winners and losers. It is critical then to think about how to deal with those who are adversely affected, how to deal with the fairness issues that are associated with the pricing reform. Let me repeat, these impacts are not unique to energy price reforms; you have losers under a regulatory program and you have losers under an emissions trading program. A big difference is that under an energy tax strategy you have the ability to deal directly with some of those issues.

For example, energy taxes are typically felt to hit lower income groups more than upper income groups, and therefore to be regressive and in that sense unfair. Energy taxes can also have differential effects by region and by industry. You can also place this in the global context, because you have the same sorts of winners and losers being created by redistributions in wealth globally. The trick then is to think of how you can compensate those who are adversely affected rather than ignore them or give up on price reform. Let me give you some examples of the magnitudes we may be talking about. Table 4 shows one set of estimates of the increases in energy-related expenditures from a carbon tax. They rise up to $322 for families earning over $50,000 a year or more. At first guess it appears that the higher income groups pay more, but as a percentage of their income they do not. These estimates are suggestions of how you can compensate by income class.

Figure 2 shows estimated differential impacts of a carbon tax by region of the U.S. It shows a big difference between the Pacific region and the rest of the country, and lesser differences among the others. It shows that some areas of the country may end up paying more than others under a carbon tax. This is part of the information you need in a tax strategy. Again, you could define similar numbers globally or nationally for other countries.

Finally, a carbon tax can create differing income or wealth effects on industry. In Figure 3 is one set of estimates from an econometric model of what happens under a carbon tax. You see output falling in a number of different industries and you can debate the size and effect of these impacts, but no matter what, the coal industry is adversely affected. A carbon tax strategy requires that you begin to think of how to compensate the coal industry.

Table 4. Estimated Increases in Energy-related Expenditures Resulting from Alternative Energy Taxes

1987 Income	$40 per Ton Carbon Tax	25¢ Gasoline Tax
Average U.S.	242	237
Less than $10,000	173	152
$10,000 - $14,999	195	179
$15,000 - $19,999	214	206
$20,000 - $24,999	237	230
$25,000 - $34,999	246	239
$35,000 - $49,999	296	298
$50,000 or more	322	318
Below 100% Poverty Line	191	179
Below 125% Poverty Line	195	183

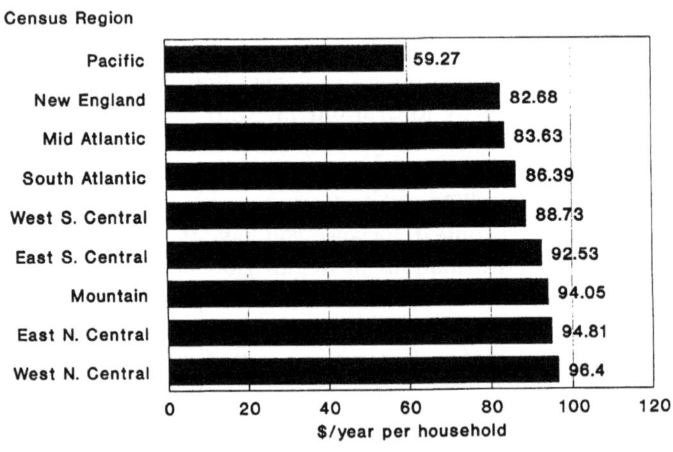

Figure 2. Changes in residential energy expenditures, based on stabilization tax.

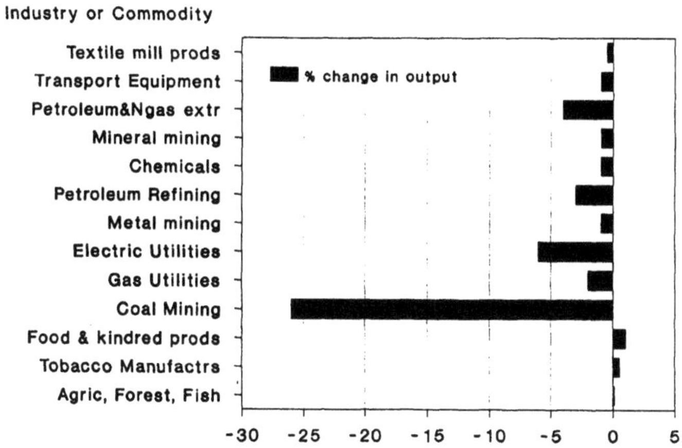

Figure 3. Effect on industrial sectors of carbon tax (stabilization tax).

Let me conclude my comments. I want to argue that price reform is a carbon reduction policy that will work. It is not the only policy; it is not the only important transition. It is a key for the transition to a less carbon-intensive energy future. It will work if you do it right. It is not necessary to give up the simplicity of a tax system, and it is not necessary to accept the burdens of a tax system. But it is necessary to think in terms of tax strategies, not just taxes.

"VIRTUAL CLIMATE" AND CLIMATE CHANGE ASSESSMENT:

PAVING THE WAY FOR WORKABLE CLIMATE CHANGE POLICIES

Charles N. Herrick

Senior Analyst
Office of the Chief Scientist
National Oceanic and Atmospheric Administration
1825 Connecticut Avenue, NW
Washington, DC 20235

Global change research has been focused predominantly on the need to resolve uncertainties surrounding physical processes that explain global change. Reducing physical science uncertainties is crucially important. However, if our goal is to guide the development of workable strategies for dealing with the societal impacts of climate change, we must also understand the interactions between climate and society.

We need to know which sectors of society are vulnerable to climate change, which relatively robust or resilient. Moreover, an improved understanding of climate-society interactions will enable us to focus research on those physical science uncertainties with the most potential to adversely affect key economic, political and cultural activities. What is needed is a climate change assessment framework that integrates physical change with societal vulnerability.

Most typically, climate change assessments characterize impacts as direct consequences of physical change (Rosenberg et al., 1990; Smith and Tirpak, 1989). For instance, crop yield change is derived using crop growth process models. Economic consequences for farmers and consumers are then estimated from changes in crop yield. Human adaptation is either not considered or viewed as an adjunct area of study. This approach (depicted below) fails to give due weight to the fact that societies already adjust for climate, and have for millennia.

Physical Change ——> Physical Effect ——> Societal Consequence

As William Clark writes, "Societies...cope with a wide range of environmental [conditions], but at different costs and within different limits. Studies of human

Global Climate Change, Edited by J.C. White
Plenum Press, New York, 1992

response to natural hazards, climate variability, and war show that both costs and limits change with time and are mixed functions of the environment itself plus the underlying demographic, organizational, and developmental characteristics of the human system..." (Clark, 1988). The fact is, we commonly offset and/or augment natural climatic conditions to support preferred patterns of life (e.g., large-scale irrigation, extensive space heating/cooling, etc.). This offset/augmentation process is significant because it may become part of the social routine, an aspect of the social, political and cultural status quo.[1]

Consider Southern California. In terms of its actual climate, Southern California is, or is nearly, a desert. However, in terms of prevalent human activities, the region is far from arid. The state has the world's sixth largest GNP, with over three-quarters of its development situated in the South. California is the nation's leader in agricultural productivity. Chief crops include thirsty species such as rice, cotton and alfalfa. Swimming pools and lush landscaping serve to accentuate Southern California's ability to augment and offset its natural climate. By all reckoning, the climate of Southern California is incongruent with its state of agricultural and economic development. However, California has created one of the most extensive water acquisition, storage and distribution systems in the world, enabling social and economic activities that otherwise would not exist.

I refer to this offset/augmentation process as "virtual climate" because we modify natural factors and create conditions more suitable for social activities. Considered from the perspective of human activity, we have virtually adjusted the climate.[2]

Virtual climate involves the large-scale, protracted and routinized augmentation/offset of natural climatic conditions. It integrates four things: (1) regional-scale physical climate factors, (2) the presence and characteristics of augmenting/offsetting technologies, (3) regional economic development and/or the socio-cultural disposition to invest in climate offset as opposed to other investments, and (4) a stable political-administrative system capable of long-term management of regional-scale climate augmentation/offset.

An assessment process designed around the construct of virtual climate would thus focus on the superimposition of climate change on those socio-cultural factors that determine the degree to which specific societal groups (regions) have already offset climate conditions to support preferred patterns of life.

Sound assessments must span many disciplines, not merely the physical and biological sciences, but the social and historical sciences as well. For example, a land-use monitoring system must be sensitive to the biological and physical characteristics of a particular land use (species diversity, albedo, percolation rate). However, it must also allow characterization of factors such as economic function of the land use, its political regulation and/or subsidization, and its cultural meaning (Stern et al., 1992). I am not prepared to argue that virtual climate can be neatly quantified or summarized in terms of one or two indices. I am, however, prepared to argue that without front-end guidance from social scientists, historians and other scholars of human development regarding the kind, scale and resolution of information needed to assess important aspects of social vulnerability to climate change, physical scientists will be unable to focus on those aspects of change which are most important to society.

As the California example suggests, the climate may not necessarily impose any one state of affairs over another. This is why the extent of climate augmentation/offset must be considered as a principal element of climate change assessment. Indeed, if climate adjustment has become an important and routinized aspect of particular patterns of socio-economic life, then social and political pressures for its maintenance will be very high, possibly as high as so-called "entitlements" such as social security and health care. An adequate methodology for the assessment of climate change impacts needs to reflect this fundamental interaction of social and physical factors.

Focusing on virtual climate serves to integrate social and physical assessment activities in a way that effectively communicates the importance of adaptation as an ongoing process. This, combined with a regional perspective, increases the chances of producing relevant information and designing workable policies.

NOTES

[1]Global change could trigger massive disruption of natural environments. The adaptability of natural ecosystems is likely more problematic than that of social systems. This essay does not address the issue of natural systems, nor does it address the reliance of social systems on ecosystems for waste dispersal, raw materials, and so on.

[2]To be clear, virtual climate is not a normative concept. I am not suggesting that adjustment is good or bad, merely that it happens. I use virtual climate strictly in a descriptive sense.

REFERENCES

Clark, William C., 1988, The human dimensions of global environmental change, pp. 134-200, *in* "Toward an Understanding of Global Change: Initial Priorities for U.S. Contributions to the International Geosphere-Biosphere Programme," Committee on Global Environmental Change, ed. National Academy Press, Washington, DC.

Stern, Paul C., Young, Oran, and Druckman, Daniel, eds. p. 142, 1992, "Global Environmental Change: Understanding the Human Dimensions," National Academy Press, Washington, DC.

Rosenberg, Norman J. et al., 1990, "Processes for Identifying Regional Influences of and Responses to Increasing Atmospheric CO_2 and Climate Change," Pacific Northwest Laboratory, Richland, Washington.

Smith, J.B., and Tirpak, D., eds., 1989, "The Potential Effects of Global Climate Change on the United States," United States Environmental Protection Agency, Office of Policy, Planning and Evaluation, Washington, DC.

EXAMINING THE LINKAGES—WHICH POLICIES WILL WORK?

Toufiq Ali

Minister
Embassy of Bangladesh
2201 Wisconsin Avenue, NW #300
Washington, DC 20007

I welcome the opportunity to participate in your eighth Annual Conference on a subject of vital importance—Global Climate Change: Linking Energy, Environment, Economy and Ethics. It provides an opening to me to discuss some issues involved and to present the developing countries' perspectives, as seen from Bangladesh.

I have been asked to speak on the subject: "Examining the Linkages—Which Policies Will Work?" May I, at the outset, frankly admit that if I knew the answers to this question, I would most probably not be here. Instead I would concentrate on implementing those policies. The case here is somewhat akin to that of development. Over the years we have learned from our experience, and are aware of a large number of policies and programs that can help in the development process. We know what has worked in some countries, at certain times, under certain circumstances. However, to apply those same policies in another environment is a risky proposition—we do not know in advance what will actually work. In the case of climate change, the situation is somewhat similar, with the proviso that we do not have the benefit of very much past experience to draw upon. Therefore, any comments that I make should probably be viewed from that perspective.

Scientists warn us that the pattern of human activity since the industrial revolution has upset the delicate balance between man and nature in the atmosphere. Anthropogenic emissions of carbon dioxide and other greenhouse gases have increased at a pace far beyond the capability of nature to absorb and recycle them. Many now fear that, unless we act immediately and in unison, the climate of our planet Earth will soon become unfit for human activity as we know it today.

Although many doubt the validity of such doomsday forecasts, most agree that there is need for precautionary measures. The very topic for this Conference emphasizes the importance you attach to it. Many fear that it may not be enough to simply halt the rate of growth of anthropogenic emissions of greenhouse gases. Our past actions may already have set the process of warming in motion; to be effective, we now must actually reduce such emissions. The United Nations was obviously seized with such concerns when it decided last year to launch the process of negotiations for a Framework Convention on Climate Change. Personally, I would regard these negotiations as being among the most important to be conducted during the current decade.

The very manner in which we consume energy, heat our homes, drive our cars, and derive the benefits of industrialization will be called into question. There are societies that consume several times more energy per capita than others at the same level of development. In the process they draw upon our common resources, much more than others do. Are such life styles to be condoned or permitted? Can some pursue their existing styles, while others be prevented from achieving the basic minimum required for a decent life? These are fundamental questions we need to address.

It is practically impossible to segregate issues linking energy, environment, economy and ethics. For instance, the use of fossil fuel or biomass for energy is a major source of atmospheric pollution. While the former is nonrenewable, the latter can be partially regenerated. Solar energy is perhaps the most environmentally friendly, but requires further development. However, many societies would accept nuclear energy as the least-cost alternative. We find a conflict. To produce energy, what is cost-effective from standard accounting, may not be the best from the standpoint of environment protection, and may raise ethical issues.

As we view it, a mix of standard textbook economics and climate issues produces a dilemma. We normally price an item by what the market would fetch for it; i.e., we expect the price to be revealed by the market. How should we price the global climate? We know that without an appropriate climate we would not be able to sustain our economic activity in its present form. Would this mean that our climate is worth US$21.2 trillion, which is the value of the world's entire GNP? With a few more complications lumped in, this becomes a much more complex case.

Take, for instance, a proposal for spending some public-sector money on an environmental project. Normally, one would carry out cost-benefit analysis, evaluate the rate of return and determine whether the project is financially viable. Unfortunately, none of the "environmental" projects would fit this bill. The rate of return is frequently "indeterminate," and the gestation period would make many a project a nightmare in a strict accounting sense. Does this mean that we should not consider environmental projects? What this does imply is that we need to develop techniques of analysis that are able to incorporate such concerns.

A related issue is that of "timeframe." When we consider an environmental project, the returns are expected at some time in the future. In other words, our current generation will be spending today for a stream of benefits that may, or may not, accrue to us during our lifetime. Are we prepared to invest for a future generation which may not really appreciate our gestures? What about those whom we are today depriving, by diverting funds that would otherwise have been available to them? Are we prepared to compensate them in any other way?

We live in a "global village." In many matters we are functionally interdependent. This is particularly true in matters relating to the atmosphere. When the Chernobyl nuclear disaster struck, the impact was felt far beyond the borders of the then Soviet Union. What is done today in the United States to reduce the emissions of carbon dioxide will certainly also help people living in other countries. This awareness of our dependence on one another is a fundamental transformation in our way of thinking.

Associated with this concept of a common heritage is the question of who should take responsibility for its protection. The answer to this is only gradually evolving, as the many conventions, treaties and protocols relating to the environment are being written and codified. Many of the answers are not so obvious as they may seem at first glance. May I place one such question for your consideration. We are all trying to negotiate a framework convention on climate change. Several countries are making commitments to reduce their emissions of greenhouse gases and help clear up the atmosphere. All countries benefit from this. Yet, what should we do about "free riders"—countries that decide not to make any contribution towards cleaning the atmosphere, but reap all the benefits from the efforts of others?

At this stage, I should digress somewhat and refer to another linkage we consider important—that among institutions in society. Protecting the global climate is the concern of all. In developing countries, many look upon the governments to provide the leadership. Yet, in cases where there are conflicting claims on their resources, there is no option but to devote attention to the immediate needs of survival. NGOs and voluntary agencies can play a vital role both in creating awareness among the people and in carrying out research in areas where governmental activity is inadequate. International agencies and intergovernmental bodies could act as clearinghouses for exchange of information, assist in the development and transfer of technology, and help develop expertise in developing countries. Other segments of society also have a role to play. Economists, for instance, could help develop appropriate tools to analyze the economics of climate-related issues. Scientists can help by carrying out research and providing us more accurate answers to some questions related to climate change. On climate change, perhaps more than in any other single issue, there is a need for a symbiotic relationship among all these groups and institutions.

Policies that are adopted, either nationally or globally, should flow from an awareness of these and other issues. Many feel that governments have to provide the leadership. A policy that is talked about and, in my opinion, is vital is an appropriate pricing policy for energy. Adequate emphasis needs to be placed on the development of environmentally friendly and renewable sources of energy. This is not easy, but our policies should facilitate this process. Issues related to the process of development need much greater attention. Most developing countries feel that only the developed countries have the know-how and the resources to develop such technologies. And, means must be found to transfer such technology to the developing countries.

Some question the need for such policies, particularly in view of uncertainty surrounding many climate change issues. We feel that many of these policies are justified in their own right, regardless of climate change. Natural resources must not be undervalued. They are exhaustible, and society has to place an appropriate valuation on them. Furthermore, inappropriate pricing leads to improper use, and with considerable associated pollution. Suitable pricing will also lead to more efficient use of our resources.

Before I conclude, I should refer to another aspect of equity. We are frequently reminded that the principal responsibility for the anthropogenic emissions of greenhouse gases lies with the developed world. It has been estimated that over three-quarters of such emissions are as a result of activity in the developed countries. It is their use of fossil fuel since the industrial revolution which has led to the problem we are confronted with. It may be argued that developed countries, given their state of knowledge in the past, did not anticipate the undesirable consequences of their policies. While apportioning blame for past mistakes may not be an appropriate guide for future action, an analysis of the past is always a sobering basis for reflection and responsible behavior in the future. What is necessary, therefore, is a common effort by all of us to meet this shared threat, with the developed countries playing a major role in protecting the global climate, and assisting generously the developing countries in coping with this problem.

It is also often said that, in the not-too-distant future, the contribution of the developing countries to the emissions of greenhouse gases would exceed that of the developed countries. We should put such generalizations in their proper perspective, lest their implications be misinterpreted. The projections of current trends may indeed lead to such results, which is inevitable and unavoidable, given our present state of technology. Furthermore, this is necessary in order to meet the requirements of growth and development in the Third World—an objective which I fervently hope the developed countries and the international community commonly share with us.

A CONGRESSIONAL VIEW OF WORKABLE LINKAGES

John D. Dingell[1]

House Energy and Commerce Committee
United States Congress
2125 Rayburn Building
Washington, DC 20515

Speaking as a veteran of two decades of energy policy wars, I would offer a few observations as we embark on yet another attempt to write a national energy strategy. The first is that these debates have always and only occurred in the wake of some sort of crisis. In the early and late 1970s, we were busily turning down our thermostats and waiting in line at gas pumps, following the Arab and Iranian oil cutoffs. This year's debate is conducted in the shadow of the Persian Gulf War.

As might be expected, a crisis atmosphere does not readily produce a calm, rational or balanced debate. The louder, more extreme and more rigidly ideological voices can dominate. With respect to energy policy, on one side are those who believe no wilderness is too pristine, no area too remote, no rock too hard or well too deep to drill, mine or exploit so that we can continue our profligate ways. On the other side are those who believe we should all ride bicycles to work and spend our evenings curled up by our wood stoves, wrapped in blankets made exclusively of natural fibers. Of course, I'm caricaturing the two points of view—but not by much.

We have tended, over the last two decades, to look at energy through the relatively narrow framework of production and conservation. Past energy policies have used a little of both. In the first energy crisis, President Nixon proposed to enlist the two to achieve "energy independence." Subsequent national energy strategies have been variations on the energy independence theme—with the exception of President Reagan's policy of cheap gasoline. Our travails and our limited success in the energy field in the 1970s, 1980s, and now the 1990s should tell us that the war for energy independence can't be won, and that the costs of pursuing it often exceed the benefits. Most energy analysts believe that during the next decade we would not reduce our dependence on foreign oil even if we adopted the most draconian command and control policies politically palatable. We have tried fuel mandates, price and allocation controls, synthetic fuels, and more—and they have all failed.

Global Climate Change, Edited by J.C. White
Plenum Press, New York, 1992

I believe that a successful national energy strategy should be based on the following strategies:

(1) We should make it easier to move from one form of energy to another;
(2) We should accept and facilitate the move toward a cleaner environment and do it in a way that fosters economic growth;
(3) We need to protect against the effects of inevitable energy disruptions; and
(4) To reach those goals, we need to use market-oriented policies.

When we talk about making it easier to switch from one form of energy to another, we are explicitly abandoning the "Project Independence" strategy. We will continue to need energy, from a wide range of sources, but we should not waste economic resources in a futile search for energy independence. Subsidies for increased synthetic fuel production succeeded only in frittering away billions of the public's money.

One of the most successful energy policies—and unexpectedly so—has been the Public Utility Regulatory Policies Act, or PURPA. Its requirement that electric utilities buy alternative sources of power at market prices has opened whole new markets. PURPA has been one of the few policies which has helped the transition to new energy supplies and uses.

Another way of thinking about this problem is to remember what happened to the horse and carriage at the end of the last century, and our relatively rapid transition to horseless carriages at the beginning of this century. The solution to the many problems of horses and carriages was not to conserve the use of horses while increasing their production. Nor was it to make carriages smaller. The answer was the transition to another technology, one which allowed increased mobility at a lower cost in a more environmentally beneficial manner. As we approach the end of this century, our continued use of oil is about as inevitable as the use of horses a century ago. Will we continue to need oil? Of course, just as we continued to need horses. But over time, we can make the transition.

The second strategy is to recognize both the public sentiment for, and the desirability of, a cleaner environment. The challenge is to do so in a way that fosters economic growth and jobs. I reject the notion that economic growth inevitably extracts an environmental penalty; conversely, environmental progress need not always come at the expense of growth or jobs.

That is how we should approach global warming. If the effects of global warming will be even half as bad as some scientists believe, we could be facing one of the most serious environmental problems in history. But the solution does not lie in shutting down economic activity; that cure would be worse than the disease. Even if the U.S. economy ceased all activity tomorrow, the production of greenhouse gases by other countries would be cause for concern.

The implications of global warming for energy policy are obvious. They mean making transitions consistent with the need to reduce greenhouse gas emissions *and* increase jobs. Some of the solutions being advanced would likely do neither. Many have suggested that excessively increased CAFE standards are an appropriate environmental and energy policy. Others believe that we should mandate the purchase of alternative-fuel vehicles. I disagree with both—not because of the ends, a cleaner environment and less oil use—but because of the means. Both approaches would have a negative economic effect. Both would probably be counter-productive. Each marginal increase in fuel economy becomes more expensive and less effective in reducing consumption. Mandating the use of alternative-fuel vehicles not ready for the market could lead consumers to reject the option. Remember some of our experiences with diesel vehicles in the mid-1980s.

As we have seen, we have doubled fuel economy standards over the last 18 years, but we are consuming more gasoline than ever. The reason is simple. In real terms, gasoline is as cheap now as it was in 1947. Gasoline is cheaper than any liquid you can purchase in the supermarket—including bottled water. As long as gasoline is cheap, and as long as we have no incentive to conserve, we will continue to use more of it.

The correct economic policy is to increase energy taxes and rebate the revenue—in other words, tax energy not people. This would both reduce energy consumption and give a sizeable incentive for the market to develop environmentally sound technologies. Of course, I am well aware that sound energy and environmental policy is not always sound politics.

A second-best policy would be to try to facilitate the transition to cleaner vehicles. From a global warming basis, biomass-based alcohol fuels and electric vehicles show the most promise. With alcohol fuels, scaling up the genetically altered microbe digesters is the main barrier. For electric vehicles, the hurdle is developing a smaller, lighter battery which can hold more energy. Tens of millions spent in these areas would reap a far greater return than the tens of billions which would be spent to comply with increased CAFE or alternative-fuel mandates. Our energy needs would be better met, our environmental goals better served, and we would open new avenues for economic growth.

The third strategy should be to recognize and establish policies which will deal with the next oil disruption—and there *will* be another disruption as long as the cheapest energy in the world comes from a politically unstable region. The Strategic Petroleum Reserve (SPR) is the only tool we have to deal with disruptions. Increasing the amount of oil we hold in the SPR is crucial. If the President had utilized the SPR at the onset of the Iraq crisis, we could have reduced the economic impact, the price spike, of the disruption—an impact we are still feeling.

Finally, our policies should be market-oriented. Markets are enormously effective in changing our patterns of consumption. The policies I have talked about use the market to bring about change. The command and control policies of regulatory mandates are of limited effectiveness. But remember: it is also true that unconstrained free markets can create problems. Many confuse market-oriented policies as an end, instead of a means to an end. Well crafted market-oriented policies, taking into account the limits of the market, will be the most effective energy policies in the long term.

For the most part, I believe the Subcommittee on Energy and Power under the able leadership of Chairman Phil Sharp has produced legislation consistent with my view of a successful national energy strategy. I hope and expect that the Congress will pass and the President will sign a successful national energy strategy—one that can sustain us well into the next century.

NOTES

[1]Presented by Michael Woo, Senior Staff Member, House Energy and Commerce Committee, United States Congress.

DISCUSSION

EXAMINING THE LINKAGES: WHICH POLICIES WILL WORK?

RUSSELL: I thought it was extremely interesting that Roger Dower was raising the question of the cost of transitions, and what we heard from Chuck Herrick was that some of the transitions we have thought of in the past as being natural or mediated elsewhere are now being taken as natural rights or human rights. So the virtual climate that we're talking about becomes something more than merely a construct; it becomes one of the constraints as far as policy is concerned.

QUESTION: Is their regressive nature the major inhibition against the imposition of carbon and energy taxes? How can you change the view of the conservative nature of governments in North America and most of the First World? Don't you need to wait for a strong political swing to the left?

DOWER: To the first part of your question, I hope not. But there are ways of looking at these questions so that they may be viewed as neither a right nor a left issue. If you structure a set of tax offsets which will provide a tradeoff with some tax offsets on the capital or conservative side, and at the same time deal with the regressive nature of the tax, then I think you have a tax that isn't simply micro-managed in terms of its politics, that has a broader appeal than a tax with one set of impacts.

Part of the strategy is to structure the tax in such a way that you balance off a lot of these factors. In the longer term, climate change is forcing us to move, and the international negotiating process is forcing us to move towards addressing these questions more rapidly than I think we would under normal political circumstances. I think the trend may be going in a better direction for us.

QUESTION: How does the U.S. global change research program part entitled "Human Dimensions of Climate Change" fit in with the discussion of virtual climate as a baseline for the assessment of climate impacts?

HERRICK: I think the human dimensions of climate change element of the U.S. global climate change program is developing. It's looking at important interactions, and it's beginning to deal with these kinds of issues that I'm raising. I've described a concept that is kind of pie-in-the-sky. We're not ready to perfectly answer it yet, but I think that's the direction that the program is going.

QUESTION: Despite political pressures to maintain augmentation or offset, like Southern California, how do we avoid the evidence that these virtual climates are or are becoming unsustainable? Should there be a socioeconomic right to continue irrigated farming in the California desert? Should this virtual climate be maintained, even though it may become more difficult and expensive to do so?

HERRICK: I'm not saying that maintaining Southern California is good or bad. I'm saying that when we do research, it's wise to look at things as they are.

When you deal with questions about how to change things or why we are avoiding certain things that seem to be stark realities, like sustainability or unsustainability, you are getting into the value side of things; you're getting into a purely political question.

How do we avoid the issue of sustainability? I thought that was something that Dale Jamieson and Doug MacLean briefed very well. With respect to answering analytical questions about this change and what we do about it, I maintain that we have to look at these systems as they are absolutely, or we will be asking the wrong questions.

DOWER: One might argue that existing water subsidy programs in California encourage the increased reliance on water-demanding crops or increased reliance on certain types of socioeconomic structures. These programs increase the cost of climate change, but can be factored quite directly into the costs and benefits of the water subsidy program. To the extent that we measure them, they become another part of a basis for promoting reforms that are basically no-regrets reforms in those areas. There are ways to incorporate all this.

QUESTION: I think an energy tax or carbon tax is a certainty in our near future—in three to five years. Given the upcoming election in '92, how do you gauge the mood of the electorate? Does the candidate supporting a carbon tax have a chance?

DOWER: I think it is true that we will see, in the industrialized countries over the next three to five years, some type of carbon tax structure. We have that throughout Europe, or gaining hold in Europe already. The European Community has proposed a combination energy carbon tax that would be adopted by all the EC countries. That's under debate now. It's not a given; it's not going to happen next year, but it's underway and it's part of the national debate.

Carbon taxes aren't yet a part of the national debate in this country. New tax policies and tax avenues are a subject of political debate. As we begin to look at alternative tax streams in this country, to fund alternative programs, to deal with deficit reduction, or to deal with our environmental problems, taxes like carbon taxes and other pollution taxes are going to take hold and be viewed as far preferable to the alternative. Two years ago, in the budget summit, there were energy taxes and a carbon tax as part of the discussion. It didn't get accepted, but that's a big change from two years before that when they wouldn't even have been part of the package.

I see movement in the right direction. It's hard to predict how and when, and I doubt very much in '92 any particular candidate is going to win the election on the basis of adopting a carbon tax policy.

DOWER: The coal industry is a loser in the carbon tax scheme, as we saw. How do you compensate this loser? The real losers are the states in which the coal industry is located. There are three states that constitute almost 50 percent of the coal production in this country: Wyoming, West Virginia and Kentucky. Many areas and many counties of those states would have nothing else, were it not for coal, and I see no other way to compensate these regions for the loss of wealth than simply large block grant transfers. The money to compensate for the lost wealth is in the range of several billion dollars. But you talk about a carbon tax that's raising $35 or $40 billion at a low stabilization level. So you have revenues to do that, and my guess is, some sort of a block grant to the states is the way to deal with compensation.

QUESTION: Nuclear power advocates still evade the question of radioactive waste. How can anyone promote an energy source that produces something that has the potential of poisoning much of the planet for hundreds of thousands of years?

BODANSKY: If I agreed with the premise of the last sentence, I would say it would be an absolute disgrace to do something that has the potential for poisoning the planet for 100,000 years. A very eloquent passage that I read about fifteen to twenty years ago on our responsibilities to future generations was written by Milton Russell, our chairman. It was an eloquent and persuasive statement, and it pointed out that you do not lightly do things that will be bad for future generations. One has a responsibility. Some of the ethical bases for that have been discussed.

Why then is there this disparity of view as to whether this does or does not create a problem? I think there is some misunderstanding as to two points: one, how the wastes are handled; and, second, what the wastes will add to our already existing radioactive world. Let me start with the first—how they're handled.

The wastes come out of a reactor in solid form, and they're kept at the reactor at present. There were two plans. In France, they turn the wastes into a glass and bury them as a glass. The present U.S. plan is to put the canisters containing solid waste into even better protected canisters. These are made in different degrees of thickness, of different materials, and these well-built containers of some thickness and some corrosion integrity are put into the driest site you can find, protected from water. These are supposed to last for 10,000 years, according to present planning. A given repository should not cause more than one death per decade. I don't want to be in the position of advocating a single death, but we're not talking about a global catastrophe; we're talking about very small additions to hazard by present standards.

To get to the radioactive nature of the earth, the whole crust of the earth is highly radioactive by some standards. That is, it has uranium, thorium, radium, potassium and so forth, radioactive isotopes that come from the original processes by which the universe was formed, nuclear synthesis. The incremental addition to that radioactivity of the earth from a waste disposal program is very small, if the wastes decay undisturbed for about 500 to 1,000 years. That's the period that one might call "acute danger."

It is true that you have a localized danger, and most future scenarios are intrusion scenarios; an unknowing person drills into the site and hits a very hot canister of radioactive waste. That could be tragic for the people involved, and you try to find a site that will have the least temptation or possibility for that. But one is talking about a very localized thing, not poisoning much of the planet for hundreds of thousands of years. One is talking about a small increment to the radioactivity already in the earth's crust. There are very elaborate plans for handling it, so elaborate that the National Academy of Science in a recent report said that our plans are too elaborate. Essentially we are frozen by the desire to get prior perfection.

It is hard to prove these things, but I and many others who have looked into it think that waste disposal will not be a hazard to future generations at all, that it will be, as far as they're concerned, a very benign and harmless thing. If I thought differently, of course, I would certainly not be taking the position I do on nuclear power.

QUESTION: What role do you see for the U.N. in both the near term and the long term in the monitoring and control of its members' greenhouse gas emissions?

ALI: We have seen in the recent past a revitalization of the role of the United Nations. One has seen it during and immediately after the Gulf War. As you are aware, the U.N. has taken a lead in terms of the negotiations for a framework convention on climate change, such as setting up the Intergovernmental Panel on Climate Change. This has produced its first assessment report, and has done a lot of authoritative work on climate change matters.

However, there is a caveat. I should refer to a little problem, and the problem is as follows. In 1990, at the second world climate conference held in Geneva, the scientists met and stated that the technology exists and cost-effective means are known for reducing carbon dioxide emissions in developed countries by about 20 percent by the year 2005. But the political leaders met after the meeting of the scientists, and did not endorse the recommendations of the scientists.

We see a problem here. The United Nations has a lot of expertise, has capability, and can sometimes do a lot of good. But can it really control actions of its members in terms of the emissions of greenhouse gases? Let us say that we should wait for the framework convention on climate change that is currently being negotiated. The next session is going to be held in Geneva very soon and we are hoping that a convention will be signed at the time of onset in Brazil in June, 1992.

But until then, it's very difficult to predict. I do believe that, in these negotiations, we have seen a great deal of desire from many countries to come to some kind of an understanding and agreement on this issue.

WOO: I look at the nuclear waste issue from a different perspective. Regardless of whether we have any new nuclear reactors, there is an intergenerational responsibility for us to deal with the waste issue, since we created it. In the federal government, I've seen a very strong desire not to deal with the waste issue.

QUESTION: Should we believe that the Corporate Average Fuel Economy (CAFE) standards are causing increased unemployment? Also, would increasing the current CAFE standards help from an energy perspective?

WOO: CAFE is essentially a supply-driven policy which is supposed to affect demand. That is, you're not giving the incentive for somebody to buy a more fuel-efficient car; you're requiring the industry to produce more fuel-efficient cars, and hopefully those will sell.

When you have these types of policies, basically you have no incentive in the system. As a result, you can still end up with higher gasoline consumption than you did when you began. But what you can succeed in doing is affecting the fate of the industry. You change the desirability that people have for certain cars, and you force industry down technological roads. Look at the demand for diesel cars, potentially a very efficient technology, but very unsuccessful. Billions of dollars were spent in research and development, which basically came to nothing. You cannot go down too many such wrong roads without an effect upon the viability of the people who are producing those commodities.

So the answer is yes, it did affect employment in places which produced automobiles. Are there policies which could improve fuel economy or fuel use in the transportation section, and result in decreased use of fuels? I think the answer is yes. Is CAFE the best one to do that? The answer I would say is probably no.

QUESTION: Isn't Sweden committed to phasing out nuclear power by the year 2000? Have they set a stabilization goal? How are they doing both?

BODANSKY: Sweden had, before Chernobyl, agreed to phase out nuclear completely by 2010. After Chernobyl, they decided to start the phasing out in 1995, predicated on finding substitutes. Now, with greenhouse effects looming larger and acid rain still there, they have backed off their 1995 phaseout. Sweden's phaseout plans are totally up in the air. There is a widespread desire to phase out, but they have not found the alternatives.

QUESTION: Isn't part of the difference in success of nuclear and solar due to differences in funding for research and differences in hidden subsidies? Can nuclear continue without a subsidy?

BODANSKY: I think it would have probably been very hard for nuclear power to get started without a subsidy, and I think the subsidy, which came indirectly from the military program, is a curse around its neck, in a way. But now, as far as I'm aware, there is no net subsidy. It was mentioned in one of the questions about price, there is a Federal subsidy to the industry. For waste disposal, the nuclear industry contributes modestly. There's a three billion dollar fund more or less sequestered by the federal government for nuclear waste disposal, but they're not spending it on nuclear waste disposal, because of Gram-Rudman difficulties.

There is a large prior investment in nuclear power. The question is, do you throw away that investment to even the playing field? I think that doesn't make sense. The other question is, do you put subsidies into solar power to advance its chances? I'd say absolutely yes. The scale of money needed isn't very large to make an enormous difference, like a billion a year. You may think I'm generous in saying that isn't very much, but it is not. Yes, I'd say you certainly should use subsidies, and time will tell what solar will do, given better subsidies.

QUESTION: If developed countries have a responsibility to help developing countries pay for rationed costs of responding to the climate threat, do developing countries also have an obligation to reduce population growth rates?

ALI: There are two parts to this question. The first part is the responsibility of the developed countries to assist developing countries, and the second part is the obligation of the developing countries. Let us analyze the two parts separately. Where does this equitable responsibility come in? In the process of development of the now developed countries, they did not worry too much about the damage being done to the environment. But now that they have developed, they have the resources, the know-how and the ability to cope with this issue. Here is where the developing countries are at a tremendous disadvantage, because they are now just trying to develop. Do we spend money on trying to develop the technology to allow energy use in an environmentally friendly way? Or do we spend money on sheer survival, on allowing our people to live? These are the very difficult choices for developing countries such as ours. Why? Because now there is a concern about damage to the atmosphere and to the global climate.

When we sometimes in international negotiations point to the responsibility of the developed countries, it is a dual kind of responsibility. It's a responsibility because of the historic nature in which the problem has evolved, and it's a responsibility because of their present contribution to greenhouse gas emissions.

Let me come to the second part of the question, that is, obligation to reduce population growth rate. There is definitely a relationship between the two areas, because when the population increases energy use also increases along with development. There will be greater pollution unless we can change the technology. But as you have seen from one of the charts presented by David Bodansky, Bangladesh has really a remarkably low figure as compared to the United States.

With two percent of the world's population, we contribute 0.06 percent of the emissions of carbon dioxide into the atmosphere. I believe we have a lot of latitude in terms of catching up.

Having said that, I would like to refer to our current governmental policies. We are trying desperately to reduce our population growth rates. It's not an easy matter. Our literacy levels are low; our health facilities are inadequate. We have all kinds of sociological problems, and our resource base is inadequate to meet all our needs. But still, population control is a topmost priority agenda of our government. Unfortunately, we have not been able to use the existing technology in the field of population control as well as we could have. But it is an issue in which we are taking a lot of interest, and we are trying to do as much as we possibly can.

But as for making one conditional to the other, such as assistance from the developed countries to the developing countries conditional to addressing population growth rates, I believe this domino kind of situation really does not work in international diplomacy. It may work in a very limited case, but two countries and societies may have totally different perceptions of the linkages between issues. We see this time and again in international relations.

QUESTION: Will the repeal of PUHCA and the opening of energy supply options result in a more environmentally benign energy policy?

WOO: PUHCA is the Public Utility Holding Company Act. It limits the ability of non-utilities to participate in the generation of power.

The answer is, the reform of PUHCA and making the electric utility system more flexible and more competitive and therefore more efficient would likely result in a more environmentally benign energy policy, as well as a more economically sound energy policy. It's interesting to note that the bill that the committee has developed is supported by the environmental community, the consumer community and the industrial community, but is held in a bit of disdain by the electric utility industry.

DOWER: I may disagree with Michael on the impact of CAFE on the macroeconomic conditions of the American automobile industry during the 1970s, as to just how significant an effect CAFE was on that industry, relative to a number of other factors. But, on the other hand, I do agree with him in terms of the impact that it's had on the bottom-line contribution of the automobile sector in this country to the key goals of environment and national security. It has clearly not been able to stop the rise in vehicle miles traveled because the cost of driving is kept low. In Europe, they have much higher gasoline taxes. Their cars are not necessarily any more efficient but you have much lower vehicle miles traveled and fewer number of vehicles per person than in the U.S.

We did some calculations, looking at simple projections of growth in global registration of automobiles. Even if you assume a doubling in the average car efficiency and just assume current growth rates of automobiles globally, the efficiency gains cannot hold automobile emissions from the sector at the present levels. You need to reduce the number of miles traveled, and only prices will do that.

One quick comment on Dr. Ali's question earlier about the role of the U.N. A key element of the Brazil conference in 1992 will be a reaffirmation of the global role of the United Nations Environment Programme, and hopefully a financial stimulus to that program to carry out its responsibilities. We look to Rio for rejuvenation and growth of the role of the U.N. Environment Programme.

QUESTION: Can we find some creative solutions on the transportation side?

DOWER: We just passed, in this country, the Surface Transportation Act, which from my point of view represents the opportunity for fundamental change in the way transportation services are offered in this country. For the first time, a significant chunk of money is going to mass transit, $32 billion.

More importantly, states that before were constrained in using funds for mass transit now have many more options available to them. We are beginning to see the potential to really change the way we commute and transport from point A to point B in this country. It's a much overlooked act, but the action goes in large part to the states, and in terms of where attention must be focused.

BODANSKY: I think education is extremely important. I don't think it's easy. I think the issues cover a very wide range. They're complex and sometimes even controversial. Education cannot be done quickly and superficially. It requires fairly ambitious programs, taking a lot of investment of time and effort.

THINKING ABOUT GLOBAL CLIMATE CHANGE

Milton Russell

Professor of Economics and Senior Fellow
Energy, Environment and Resources Center
The University of Tennessee, Knoxville
Collaborating Scientist
Oak Ridge National Laboratory

Climate change is a world-class problem in both meanings of that term. The "Global Climate Change: Linking Energy, Environment, Economy and Equity" linkages that are the subject of this book demonstrate the global reach of the effects, and their complexity challenges the most nimble of minds and the deepest of intellects. No short chapter could summarize these linkages, even if they were all clear. And no chapter of any length could interpret and define these linkages because so far they defy clear articulation.

We simply do not know enough. There are chasm-sized gaps in scientific understanding and measurement. Beyond that, there are no road maps to indicate climate response to possible behavioral shifts. Finally, divisions on how to value consequences and on how they should be distributed among the peoples of the world and among generations add further layers of opacity that confound understanding. There is no simple or straightforward answer to where we want to be from the perspective of where we are, and no shared vision of how to get there.

In this context, to even summarize the ideas, concepts, views, concerns and values expressed in the preceding chapters is both futile and pretentious. To presume to draw universal conclusions about the substance and direction of the linkages is worse. This is a subject in flux that lends itself to interpretations as numerous as those who reflect seriously upon them. The preceding chapters will induce some shift in those interpretations by those who peruse them. Perhaps, just perhaps, the diversity of those interpretations will narrow among the readers of this volume, but the new insights offered may just as easily jolt some away from their tentative moorings and lead to greater personal uncertainty. And that is as it should be with world-class problems. When resolution gets clearer and better defined, such problems lose their luster—and the real work of meeting them head-on can begin.

Global Climate Change, Edited by J.C. White
Plenum Press, New York, 1992

That point has not really arrived in my opinion when it comes to global climate change. Many may conclude, however (as I do), that in part, and for the present, the path is sufficiently clear that it is time that movement can begin in a direction that, more likely than not, will lead to a better outcome. I don't wish to dwell on the matter here, but the "no regrets" strategy applied to both mitigation and adaptation appears to have much in its favor.

I don't choose to concentrate on such small victories, however, because important as they are, they do not do justice to the world-class problem that those who organized this conference took upon themselves. Instead, I want to concentrate on the process by which we might think about problems such as global climate change, and on how the fact of being forced to think about this problem may change the way we approach other issues in the future.

Intellectual history and its defining modes of thought form a continuum, but a continuum on which we tend to impose a sense of order that looks for turning points. It is both neat and somehow satisfying if those turning points fit comfortably with time periods that seem to have significance of their own—decades, half-centuries, centuries or millennia. I think that as we approach the 21st century, we are at such a turning point. The conference that gave rise to this book coincided with the 50th anniversary of Pearl Harbor—for the U.S. the defining moment when we turned away from inward preoccupation with the Great Depression and outward—first to survival and then to victory in World War II. Pearl Harbor launched for Americans a half-century quest for national security against human enemies faced across national boundaries.

Intellectually and emotionally, World War II left the legacy of the ability to conceive of global war in an era of weapons of mass destruction—the nuclear age. In a strange and perverse way, the threat of mutual annihilation tied the world together as it had never been tied together before. We had talked for at least two millennia about being our brother's keeper and now that reality was upon us. After 1945, we lived in a world in which each of us, in some sense, had a responsibility for the continuity of all the rest of us on the planet. And it was reciprocal; we each depended on others as perhaps never before. International peace became everybody's business because there was no place to hide, and no one was not involved. International relations for the first time became, not matters of special interest for an elite few, but matters of distinctly personal interest to everyone. In a sad and ironic way, one living legacy of World War II to the succeeding half century was the "One World" that visionaries such as Wendell Wilkie had only foreseen in rhetoric.

World War II marked another turning point as well. It came to define a massive change in the dominant culture. I ask you to recall the grand novels of C.P. Snow. He described the two intellectual traditions of science and the humanities and chronicled the clash of these cultures in the corridors of power. He recorded the ultimate ascendancy of science. I think in many respects, the impacts of science and technology on World War II broke the perceived doctrines of the liberal traditions, and brought the "new men" of science into power.

The most obvious basis for this change was the success achieved when scientists moved directly into the war effort. The power of focused research produced the radar and sonar which kept Britain in the war. It provided the proximity fuse which magnified the lethal effects of brute force weapons such as artillery. Science and technology substituted scientists and engineers for the hordes who struggled in trenches in World War I. And of course they produced nuclear weapons which ended the war and created the shield of terror that, combined with enlightened pragmatism, kept East and West from a climactic violent struggle.

For whatever reason, the new mode of thought forced on those who had power after World War II was one less tied to the past. It opened to influence new bases for making decisions. In John Dewey's terms, the instrumental values associated with technology and science achieved dominance. The value became "what works," not the ceremonial test of "what is right," based on perceived doctrine. Nowhere was this more important than in the ways in which public policy decisions were made, or at least explained and defended. Pragmatism was in the ascendance, even if sometimes only as the handmaiden of ideology.

This new way of thinking was not limited to international relations. In ways that people cared about, science worked. Very few of the societies all over the earth failed to be affected by the power of looking at problems in ways that sought solutions through open examination of alternatives. This questing mind-set tended to yield solutions that made people better off, and they recognized it. Living through this period there was a sense of continuity and natural change. Looking backward to the 1930s and beyond, there is a sense of inevitability about all this as part of an evolutionary process. But, I submit, this was not in fact the case. Instead there was an inflection point at World War II that shifted the way of thinking. First, it created a sense of connectedness because it showed we could survive only at the tolerance of others. Second, it altered the way of thinking toward the focused use of science and technology in meeting challenges.

We are now poised to enter the new millennium with this altered intellectual heritage. In my view we are today living through what will someday be seen to be another turning point of similar significance—and this has to do with global climate change. Many assertions have been made over the course of the last dozen years about environmental claims to global attention. Many will recall the Brundtland Commission report of a few years ago entitled *Our Common Future.*[1] In exalted rhetoric it spoke of how all of us were tied together by the environment and by the need to bring the blessings of a higher living standard to the poor peoples of the world through sustainable development. Before the Brundtland Commission there was the era of discussion of the North-South conundrum heralded by the Commission chaired by Willy Brandt.[2] But as a counterpoint, there was also the work of René Dubos with his injunction that it was proper to think globally, so long as one acted locally, and his view that man need not be thought of as exploiter and despoiler.[3]

In essence, most of the issues that *Our Common Future* addressed were either local or regional in character. The Brandt Commission which preceded it mostly called for one group—the developed countries—taking responsibility for others, at the developed country's expense. The issues to which attention was drawn by these and a shelf-full of other books ranged from species preservation (which except as an abstraction is unimportant to most people) to water quality (which is important only if one drinks that water). As a trite example, they included saving the whales, which is very important to some people but not particularly central to the lives of the nomads of Tibet. In short, all of these issues that gained attention in the 1970s are in one way or another rather localized or specialized, even though when taken together they may touch us all with shorter or longer chains of influences. Put differently, it requires a leap of altruism that most people don't possess to care deeply and personally about the quality of the air in a city they never expect to visit.

For this reason these calls for a global commitment, backed with short-term, discernible, personal sacrifices of real satisfactions, have touched only a narrow sliver of the public in ways that really count. There have been nods of agreement that "something ought to be done." But there have not been major and widespread behavioral changes that make a big difference. There has been no outpouring of support that would mobilize the political will to increase taxes or to deter needed national or local expenditures to meet environmental demands that are diffuse or located elsewhere. And this is not surprising.

With global climate change, as with nuclear war, there is no place to hide and everyone is responsible. This fact changes the dynamics of the intellectual and emotional processes that led only to nods of agreement when it came to a collective response to environmental problems located elsewhere. It means that perhaps the political will can be tapped to say we should do something—once the matter is clear that something should be done.

Interestingly, it appears that such an intellectual transformation may be transferable to other diffuse and indirect environmental challenges. Once the gap is crossed that says what happens elsewhere affects values of direct concern, it is easier to see personal responsibility for the preservation of species in the tropical rain forest. In that sense, the recognition of the prospect of harmful human manipulation of the climate opens the door to meaningful concern for "One World" in the environmental sense.

Global climate change may prove in the environmental sphere to be the functional equivalent of the threat of nuclear war in the geopolitical sphere. It has made us in some personal sense our brother's keepers, and, henceforth, mutual hostages to fate. Thus far I have drawn a parallel between the issue of global climate change and the geopolitical consequences of World War II that arose from collective terror. In each case, I suggest, the shift in the perception of threat caused people to lift their eyes from local concerns to consider the broader ramifications of their actions. It opened the way to a collective response. In the case of nuclear war, the threat came from across national borders, while in the case of climate change, it comes from unfettered pursuit of individual and national activities.

There is another parallel that arises as well. This is in the clash of cultures—in this case with respect to how to address climate change. The preceding chapters in this book document two approaches to where we are today with respect to climate change—the approach of the analyst and the approach of the political decision maker.

A political decision maker lives in a binary world. When the vote was called on the Clarence Thomas nomination, many Senators sought to explain their reasoning and to blur their decisions. But ultimately, they had to vote, "yea!" or they had to vote, "nay!" When the President gets a bill from the Congress he may artfully craft the message delivered with his decision, but there is no place for him to sign for "maybe"; or for "yes, sort of." The political world is made up of binary decisions. This is a world where any compromise must come in the options before the decision is made. But, ultimately, it is a world of answers, a world of absolutes, a world where "choices are made."

Analysts, on the other hand, live in a different, fuzzier world. The analyst's culture thrives on complexity. Reading between the lines on the preceding pages gives the sense that global climate change, as an issue, has all the elements that are dear to analysts, and all the elements that are anathema to responsible decision makers. There is scientific uncertainty. There is the differential effects of time. There are long time horizons. There are interactions on the physical level of a complexity and of a size that even our largest computers cannot handle. There are significant but uncertain distributional effects. There are multiple attributes affected on multiple dimensions. There are complex, but also significant, two-way linkages to other problems. Each advance in understanding gives rise to further questions, and the goal of having them all sorted out seems to recede, not grow closer. This is what makes global climate a world-class problem. Looking for answers through analysis, then, leads to paralysis—at least for now.

But what about the political approach? The political world is paralyzed as well when the binary choices available are not politically acceptable—thus the search for a formula that meets all sides of the issue. The "no regrets" strategy is one result, and not a bad approach for a period when uncertainty prevents an unambiguously wise conclusion as to what to do.

Yet, decisions will be made, including perhaps a decision to delay a final decision. In truth, it is perfectly reasonable in some circumstances to advise, "Don't act, just stand there." Whatever the decision, it will be deeply controversial because at this stage there are no unambiguously correct answers, and the presence of multiple dimensions means that a satisfying midpoint compromise is not possible.

Both the political culture and the analytical culture are involved in these clashes. The interesting thing is that the clashes come across countries, across disciplines, and across special interests, including social classes. There is not even total agreement on what is important, much less on what weights to place on each attribute and what actions should be taken. This diversity has been demonstrated in the chapters presented here; even when a surface appearance of agreement exists, when the "agreement" is probed, fissures appear.

How does this description of decisional disarray comport with my earlier comments on the issue of global climate change being a forerunner of a new era of intellectual history? The answer lies in the way this problem is being addressed. First, it is being conceived as a collective problem facing all the peoples of the world, not as one in which there are both sides and bystanders. Second, unlike the "One World" engendered by fears of nuclear disaster (which disciplined separate sides with fears of the other), this is a case where the superordinate goal of prosperity and security for all is the motivating force for cooperation. Third, the analytical complexities and scientific uncertainties are explicitly taken as part of the political decision process, not as matters to be resolved first and then, once a binary decision is taken, to be repressed. Finally, the perception of "an answer" is absent. It is being replaced by the perception that there are serial decisions to be taken on an "act and learn" basis. This is intellectually poles apart from the more common approach of "learn then act." Science, analysis and political action are not seen as sequential activities but contemporaneous ones.

This moves the intellectual process one further step from the pre-World War II model where these activities were not even within the same culture. This is the basis on which I reach the following conclusion: Attention to global climate change may mark a shift in the dominant world view just as World War II can be identified as the turning point when science transformed the political process and the everyday life of people. Science and scientists were allowed into the corridors of power when what they had to offer tempered the threat to existence of a whole system. In the same way, the complexities of global climate change today present analysts no way to fit their results into the binary world of political decisions on a national level. A new way of thinking is required. While that way of thinking is evolving, its outlines are far from clear. The linear process of learn, then decide, simply does not fit—even if the decisions could ignore the divisions of nation-states, which have themselves so far defied the eraser.

Some of the chapters in this book describe the high drama of national decision making and the even more gripping story of how those decisions transcended national boundaries. The Montreal Convention on the ozone layer is one example. The negotiations in Rio de Janeiro at the World Environment and Development

Conference offer another occasion when national interests will clash, and how this will sort out is not yet clear.

What is evident is that decisions on global climate policy will present an opportunity to examine how the old order changes, an opportunity not unlike that presented in the San Francisco conference at which the United Nations was born. Perhaps those now wrestling with the global climate issue will find themselves bit players in some future C.P. Snow novels which chronicle the rise of a new way of thinking—and of acting on global issues. What seems clear to me is that the intellectual structure for forming and executing public policy on national terms that existed in 1940 was replaced by one which placed science and analysis at the front end of decisions and action. Similarly, the current model will change as the world moves, hopefully successfully, in responding to the challenges that require a new way of thinking about how the peoples of the world will relate to each other and to the natural environment. The issue of global climate change is one forge in which this new way of thinking will be cast.

NOTES

[1]World Commission on Environment and Development. *Our Common Future*, Oxford University Press, New York (1987).

[2]W. Brandt. *The Common Crisis North-South: Cooperation for World Recovery*, Massachusetts Institute of Technology Press, Cambridge, MA (1983).

[3]R. Dubos. *The Wooing of Earth: New Perspectives on Man's Use of Nature*, Charles Scribner's Sons, New York (1980). While Dubos wrote broadly on many issues, this book is perhaps his most elegant collection of essays devoted to the relationship between man and nature.

CONFERENCE PARTICIPANTS

William Achor, Western Maryland College, Westminster, MD
Toufiq Ali, Embassy of Bangladesh, Washington, DC
David Anderson, American University, Washington, DC
Annette L. Bach, North Dakota State University, Fargo, ND
Robert Barker, Cornell University, Ithaca, NY
Adrienne Bauer, American University, Washington, DC
Lawrence Belle, Rochester Institute of Technology, Rochester, NY
Catherine Benham, US Environmental Protection Agency, Washington, DC
William Benusa, Chevron Corporation, San Francisco, CA
Nicholas Berenyi, Westvaco Corporation, Summerville, SC
Perry Bergman, US Department of Energy, Pittsburgh, PA
Margaret Best, Bermuda Biological Station, Ferry Reach, Bermuda
Jan Beyea, National Audubon Society, New York, NY
David Bodansky, University of Washington, Seattle, WA
Gwen Bonebrake, Center for Global Change, Washington, DC
Susan Bush, EOS, Washington, DC
Patrick Butler, National Organization for Women, Washington, D.C.
Diana Campbell, Lighting Design Lab, Seattle, WA
Jane Caplan, Smithsonian Institution, Washington, DC
Michelle Carlos, American University, Washington, DC
Rachel Chang, American University, Washington, DC
Edgar Chase, Fairfax County Air Pollution Division, Fairfax, VA
Paul Cicio, The Dow Chemical Company, Washington, DC
Julie Clendenin, Edison Electric Institute, Washington, DC
Doug Cogan, Investor Research Center, Plainfield, NH
Rob Coppock, National Academy of Sciences, Washington, DC
Susanna Cordts, CEI, Rochester, NY
Ken Corum, Northwest Power Planning Council, Portland, OR
Laurence Cotton, Cambridge Forum, Cambridge, MA
Alex Cristofaro, US Environmental Protection Agency, Washington, DC
Marjorie Davis, Union of Concerned Scientists, Decatur, GA
Louis Davis, Union of Concerned Scientists, Decatur, GA
Catherine Dillingham, Fairfield University, Fairfield, CT
Jerry Dion, Arizona Energy Office, Phoenix, AZ
Jack Donnan, Ontario Ministry of the Environment, Toronto Ont.
Roger Dower, World Resources Institute, Washington, DC
Giles Endicott, Ontario Ministry of the Environment, Toronto Ont.

Christine Ervin, Oregon Department of Energy, Salem, OR
Deborah Estes, Island Press, Washington, DC
Karen Evens, Minnesota Power, Duluth, MN
William Fang, Edison Electric Institute, Washington, DC
Art Fish, Jr., Argonne National Laboratory, Washington, DC
Laury Marshall-Forbes, Izaak Walton League, Arlington, VA
Elizabeth Fox, Chester, MD
Peter Fraser, Ontario Ministry of Energy, Toronto, Ontario
Robert Friedman, US Congress Office of Technology Assessment, Washington, DC
Hiroshi Fujimaki, Cornell University, Ithaca, NY
Susan Gardner, University of Wisconsin, Madison, WI
Robert Gehri, Southern Company Services, Birmingham, AL
Kristalina Georgieva, Massachusetts Institute of Technology, Cambridge, MA
T.J. Glauthier, World Wildlife Fund/The Conservation Foundation, Washington, DC
Patrice Gordon, Congressional Budget Office, Washington, DC
Howard Gruenspecht, US Department of Energy, Washington, DC
Paul Gunter, Nuclear Information Resource Service, Washington, DC
Scott Hajost, Environmental Defense Fund, Washington, DC
Lynne Carter Hanson, University of Rhode Island, Narrangansett, RI
Nelson Hay, American Gas Association, Arlington, VA
Udi Helman, CSIS, Washington, DC
Charles Herrick, National Oceanic & Atmospheric Administration, Washington, DC
Martha Hewett, Center for Energy & Urban Environment, Minneapolis, MN
Ruth Foell Hill, UNA/USA, Charlotte, NC
Edward Hillsman, Oak Ridge National Laboratory, Oak Ridge, TN
David Hodas, Widener University School of Law, Wilmington, DE
Jau-Inn Huang, Columbia University, New York, NY
Carole Huther, CEI, Rochester, NY
Dale Jamieson, University of Colorado, Boulder, CO
Russell Jones, American Petroleum Institute, Washington, DC
Joan Jordan, National Science Foundation, Washington, DC
John Justus, Library of Congress, Washington, DC
Robert Kane, Burke, VA
Joan O. King, 20/20 Vision, Sautee, GA
Gretchen Kolsrud, US Congress Office of Technology Assessment, Washington, DC
Elizabeth Kraft, League of Women Voters, Washington, DC
Carl Lankowski, American University, Washington, DC
Daniel Lashof, Natural Resources Defense Council, Washington, DC
Jeff Laws, Montgomery County Public Schools, Dickerson, MD
Scott Layne, Maryland Energy Administration, Kingsville, MD
Alice LeBlanc, Environmental Defense Fund, New York, NY
Samuel Leonard, General Motors Technical Center, Warren, MI
Robert Lindgren, Center for Global Change, College Park, MD
Francois Louis, Renault USA, Rockleigh, NJ
Douglas McLean, University of Maryland, Baltimore, MD
Frederick Mangelsdorf, Texaco, Inc., Beacon, NY
Gary Marfin, CONOCO, Inc., Houston, TX
Tony Mazzocchi, Public Health Institute, New York, NY
Robert McFadden, Motor Vehicle Manufacturers Association, Washington, DC
Barry McNutt, US Department of Energy, Washington, DC
Jodi Mendel, Concern, Inc., Washington, DC

Bert Metz, Netherlands Embassy, Washington, DC
Stanton Miller, ES & T, Washington, DC
Patricia Monahan, Madison, WI
Eric C. Mosher, Wisconsin Department of Natural Resources, Madison, WI
Frank Muller, University of Maryland, College Park, MD
Madeline Nagy, Smithsonian Institution, Washington, DC
Barbara Nash, Virginia Cooperative Extension Service, Arlington, VA
Cindy Axinn Newberg, Japan Electric Power Information Center, Washington, DC
Adele Gay Nicholson, Cornell University, Ithaca, NY
Donna Nickerson, US Environmental Protection Agency, Washington, DC
William Nitze, Alliance to Save Energy, Washington, DC
Mary Novak, DRI/McGraw-Hill, Lexington, MA
James O'Hear III, Department of Architecture & Planning, Washington, DC
Ray Oglesby, Cornell University, Ithaca, NY
Nancy Parks, Sierra Club, Aaronsburg, PA
Steve Pernick, Jr., Duquesne Light Company, Pittsburgh, PA
Susan Petronio, Sterling, VA
Elizabeth Pixley, Monroe Community College, Pittsford, NY
Anne Polansky, Solar Energy Industries Association, Washington, DC
Mark Popovich, Center for Clean Air Policy, Washington, DC
Sally Rakow, California Energy Commission, Sacramento, CA
Thomas Reichenbaugh, Montgomery County Public Schools, Kearneysville, WV
Daniel Reifsnyder, US Department of State, Washington, DC
Kenneth Richards, Bethesda, MD
Richard Richels, Electric Power Research Institute, Palo Alto, CA
Diana Furchgott Roth, American Petroleum Institute, Washington, DC
Milton Russell, Center for Energy & Environment, Knoxville, TN
Karin Santora, Washington International Energy Group, Washington, DC
Lee Schipper, Lawrence Berkeley Laboratory, Berkeley, CA
Karen Schmitt, Environmental & Energy Study Institute, Washington, DC
Leo Schrattenholzer, IIASA, Laxenburg, Austria
Hugh Schratwieser, US Department of Commerce, Washington, DC
Nita Settina, Center for Policy Alternatives, Washington, DC
Lynn Simarski, EOS, Washington, DC
Dr. Sklaroff, British Embassy, Washington, DC
Douglas K. Smith, D.K. Smith & Associates, Stoneham, MA
Nancy Paige Smith, St. Mary's College of Maryland, St. Mary's City, MD
J.R. Spradley, Jr., US Department of Commerce, Washington, DC
Marnie Stetson, World Watch Magazine, Washington, DC
Richard Stewart, Georgetown University Law Center, Washington, DC
Valerie Strauss, League of Women Voters, Washington, DC
Nicholas Sundt, Energy, Economics & Climate Change, Washington, DC
Giorgio Tesi, National Science Foundation, Washington, DC
Terry Thiele, General Electric Company, Louisville, KY
Elizabeth Thorndike, CEI, Rochester, NY
Gerald Threadgill, DRI/McGraw-Hill, Washington, DC
Craig Van Grasstek, Washington Trade Report, Washington, DC
Hans Van Zijst, Netherlands Embassy, Washington, DC
Bill Wagner, CEI, Rochester, NY
Linda Wall, CEI, Rochester, NY
Richard Wallace, Washington, DC

Barbara Wells, National Governors' Association, Washington, DC
Clas-Otto Wene, Chalmers University of Technology, Upton, NY
John Weyant, Stanford University, Redwood City, CA
Jim White CEI Rochester, NY
Michael Woo, US Congress, Washington, DC

CONFERENCE PROGRAM

Thursday, December 5, 1991

Welcome and Introduction - **Elizabeth Thorndike**

Keynote Address
RISKING CLIMATE CHANGE - ACTION IN THE FACE OF UNCERTAINTY
Raymond Oglesby, Global Environment Program, Cornell University

Session I: A SUSTAINABLE ENERGY FUTURE - TECHNOLOGICAL OPTIONS FOR CHANGE
Chair - **T.J. Glauthier,** World Wildlife Fund - The Conservation Foundation

Supply side: How far can efficiency, conservation and renewables take us in meeting global energy demands?
Robert Friedman, U.S. Congress, Office of Technology Assessment

Demand side: What are options and barriers to change in engineering, end-use applications and lifestyle?
Lee Schipper, Lawrence Berkeley Laboratory

Why do cost analyses differ? Model sensitivity to alternative assumptions.
Richard Richels, Electric Power Research Institute

Session II: INTEGRATING ENERGY, ECONOMY, AND ENVIRONMENTAL CONCERNS
Chair - **Nelson Hay,** American Gas Association

"No regrets" Tax Reform
Alex Cristofaro, Air & Energy Policy Division, U.S. Environmental Protection Agency

Multiple Benefit Environmental Policies
Daniel Lashof, Natural Resources Defense Council

Market Mechanisms
Alice LeBlanc, Environmental Defense Fund

Local/Regional Policies - Acting Globally by Thinking Locally
Edward Hillsman, Oak Ridge National Laboratory

Session III: IMPLICATIONS FOR DECISION MAKING
Chair - **Jan Beyea,** National Audubon Society

Sectoral Perspectives

Government
Howard Gruenspecht, U.S. Department of Energy
Electric Utilities
William Fang, Edison Electric Institute
Auto Industry
Samuel Leonard, General Motors Technical Center
Manufacturing
Frederick Mangelsdorf, Texaco, Inc.

Societal Considerations

Population
Gretchen Kolsrud, U.S. Congress, Office of Technology Assessment
The Public as Decision Maker
Rob Coppock, National Academy of Sciences

Friday, December 6, 1991

Session IV: STRATEGIES FOR INTERNATIONAL NEGOTIATIONS
Chair - **William Nitze,** Alliance to Save Energy

Status of Framework Convention Talks
Daniel Reifsnyder, U.S. Department of State

Equity in International Agreements
Richard Stewart, Georgetown University Law Center

Session V: THE ETHICS OF LIVING IN A GLOBAL GREENHOUSE
Chair - **William Nitze,** Alliance to Save Energy

Corporate and Personal Responsibility
Dale Jamieson, University of Colorado

Intergenerational Equity

Douglas MacLean, University of Maryland, Baltimore County

Session VI: EXAMINING THE LINKAGES - WHICH POLICIES WILL WORK?

Chair - **Milton Russell,** University of Tennessee and Oak Ridge National Laboratory

A panel of experts representing the perspectives of science, economics, society, U.S. government, international agencies and developing nations.

David Bodansky, University of Washington
Roger Dower, World Resources Institute
Charles Herrick, National Oceanic and Atmospheric Administration
Toufiq Ali, Embassy of Bangladesh
Michael Woo, House Energy & Commerce Committee, U.S. Congress

SUMMATION

Milton Russell

INDEX

Acid Rain, 57-58, 98-99
AEEI, 46, 48-49, 51
Afforestation, *see* Forests
Agriculture and Stabilization Service, 31
Alternative Energy, *see also* Biomass; Geothermal; Hydroelectric; Methane; Natural Gas; Nuclear; Research; Solar
  availability, 47
  costs, 47, 212
  consumption of encouraged, 30, 213
  education programs needed for, 67
  emissions reductions and, 15
  forecasts, 68, 109
  NE-BAK, 47 (defined), 50-51
  RNEW, 47 (defined), 48, 50, 51
  research and development needed, 25-26, 30, 66-67
Aerosols, Atmospheric, 104
Australia, 172
Automobile, *see* Motor Vehicle
Autonomous Energy Efficiency Improvements, *see* AEEI

Baker, James (U.S. Secretary of State), 164
Bangladesh, 207
Behavior and Behavior Changes, *see* Corporations; Public
Belgium, 192
Biomass, *see also* Forests
  as alternative to fossil fuel, 17, 192, 213
  carbon taxes and, 94
  ethanol from, 47
  protection of forests and, 94
  use of trees and, 31, 32
Brazil, 173
Brundtland Commission, 225
Building Sector, *see also* Conservation; Policy Making, 20-23
Bureau of Land Management, 31
Bush Administration (U.S. President George Herbert Walker Bush)
  carbon dioxide emissions and, 104, 164

Bush Administration (cont'd)
  decision making and, 70
  emissions trading and, 80
  "no regrets" policy and, 164
  population policies and, 148
  reelection and environmental policies, 144, 164
  UNCED and, 166
  U.S. public opinion and, 144
  values and, 184
Business Council for Sustainable Development, 165

CAFE Standards
  fuel consumption and, 75, 112, 121
  fuel costs and, 75
  increases in, 66, 78, 212-213
  motor vehicle industry and, 111-112, 114, 220
  unemployment and, 218
California
  effects of climate change and, 204
  energy efficiency and, 55
  health of economy and strictness of environmental laws, 147
Campbell, Joseph, 183
Capitalism, and ethics, 171, 184
Carbon Dioxide, *see* Emissions; Emissions Reductions; Emissions Trading; Greenhouse Gases; Offsets
Carbon Dioxide Offset Policy Efficiency Act of 1991, 78-80
Carbon Emissions, *see* Emissions
Carbon Taxes, *see* Taxes
CFCs, *see* Chlorofluorocarbons
Chlorofluorocarbons (CFCs), 62, 87, 158-159, 162
Chernobyl, effect of accident on global environment, 208
China, and coal, 109, 121
Clean Air Act, 103
  acid rain and, 18, 79-80
  emissions reductions and, 69-70, 79-80

Clean Air Act (cont'd)
market forces and, 73, 163
vehicle emissions and, 53, 87
Climate (and Climate Change)
adjustment to, 203-205
Bush Administration and, 104, 144, 164, 166, 184
compared with other global threats, 104, 226
effects of, 4, 6-7, 31, 81, 82, 122, 147, 157, 172, 204
island nations and, 157, 158
need for new ethical paradigm and, 171-172, 227
oceans and, 155, 157-158
preventive measures, 161
sociological aspects of, 7, 169-86, 203-205, 215
technologies to address, 204
temperature and weather, 4, 6, 9-10, 93-94, 122, 154, 178, 189
uncertainty about, 3, 4, 6-7, 81, 116
water supply and, 6, 9
CO_2, *see* Emissions; Offsets; Emissions Reductions; Emissions Trading; Greenhouse Gases
Coal
co-firing with natural gas, 28-29
consumption
decreasing, 28
in China, 109
in former Soviet Union, 109
in OECD countries, 109
in United States, 109, 155
demand for, 28, 108-109
industry
and carbon taxes, 200, 216
economic stress and, 164
impacts of emissions reductions to, 105
Corporations
attitudes of, 88, 99, 172-174
behavior and behavior changes, 169, 172-174
ethics of, 169-174
initiatives of, 78-79, 117, 164-165, 174
public education and, 117
responsibilities of, 75, 87, 99, 171, 174
Corporate Average Fuel Economy, *see* CAFE
Costs
environmental compliance, 76, 98, 103
fuel, 39, 40-41, 75, 213

Decision Making, *see also* Policy Making
need for information in, 89
international, 85
need to consider many perspectives, 118
on various governmental levels, 81-89
Department of Energy, 28, 31, 67, 105
Developing Countries, *see* Economically Developing Countries; *see also* Industrialized Countries
Dow Chemical, 165
Dubos, Rene, 225

EC, *see* European Community
Economic Analyses, 58
Economic Development, energy consumption and, 127
Economically Developing Countries
behavior changes likely, 169
effects of climate change on, 104
emissions produced by, 154, 190, 210
emissions reductions by, 98, 105-106, 122
energy consumption increasing and, 190-191
energy efficiency and, 55
population policies and, 148, 219, 220-221
technology needs and, 161-162
Education, *see also* Public Education; Information, university level, 67
Efficiency, *see* Energy Efficiency
Electric Utilities
emissions reductions over past 2 decades, 107-109
policy planning, 28-31, 107-109
Electric Vehicles, 213
Emissions, carbon dioxide; *see also* Greenhouse Gases
economically developing countries and, 154, 210
in Eastern Europe, 43
fees and, 76
immediate cuts being made in, 132
increases in
projected, 33, 109, 117-118
historic, 191
population growth and, 118
industrialized countries and, 154
stabilization and, 57-58, 108, 165-166
taxes, *see* Taxes
Emissions Caps, U.S. opposition to, 143-144
Emissions Reductions, *see also* Alternative Fuels; Conservation; Financial Mechanisms; Offsets; Research & Development
costs
evaluation of, 105
borne by customers, 107, 118
uncertainty in estimating, 14-15
demand-side management, 20-21, 27, 121
in European countries, 132, 154
economic effects, 45-52, 54-55; 116, 195-201, 212
economically developing countries and, 98, 122
enforcement of, 163, 218
in industrialized countries, 104, 105-106, 123, 154, 189
international policies for, 121-122

Emissions Reductions (cont'd)
in Japan, 154
market forces and, 34, 43, 213
motor vehicles and, 23-25, 53, 87
natural gas and, 79, 104
nuclear energy and, 28, 193
policies, 15-34, 104, 154-156
regulatory approach, 34
soot emitted from jets and, 116
vs. stabilization, 58
supply-side policies, 16-17
technology and, 15, 16-34, 98
top-down approach, 105, 154-155
in the United States, 13-14, 57, 154-156
Emissions Trading
Clean Air Act and, 77, 89-90, 163
costs, 94-95
effectiveness of, 27, 77-80, 97
electric utilities and, 29
GNP and, 94
description of, 17-18
international, 159-160, 62-63
"polluter pays" principle and, 98
as preventive measure, 78
reductions of CO_2 and, 77-80
Energy Conservation
building codes and, 20, 21-22
economics and, 131-140, 155
energy audits, 22
inefficient appliances and, 16, 21
light bulbs, high efficiency, 20, 144-145, 149
mass transit, 23, 25, 88-89, 221
"no regrets" policy and, 116
public education about, 22-23
standard of living and, 117
travel reduction, 16, 23-25, 42, 67, 87, 88, 112-113
Energy Consumption
constraints to, 109, 130-137
in economically developing countries, 131-140, 190-191
in France, 191
in OPEC countries, 131-140
projected increases in demand, 108, 109, 133, 191
population growth and, 127-140
transportation and, 23
in U.S., 23, 131-140
U.S. and African countries compared, 130
Energy Efficiency
behavior changes and, 40
California and, 55
differing countries' views of, 55-56
economic growth and, 131-40
education (university level), 67
electric utilities and, 29, 65
emissions and, 14, 37
evaluation criteria and, 56, 95
households and, 39
improvements in, 37-40, 213
Energy Efficiency (cont'd)
manufacturing and, 38
speed limits and, 39-40
state vs. federal initiatives, 55
transportation and, 39-41
U.S., 40-41, 55
Equity, 103, 157, 158-60, 177-181
Ethics (and Values)
Bush Administration and, 184
capitalism and, 171, 184
culturally based, 183-184
information needs and, 183
need for, 169-174
of past generations, 179-180
responsibility and, 179-181
science and, 224-228
Traditional Native American, 184
values, defined, 171
European Community (EC)
carbon tax and, 163, 216
compared to U.S., 70
emissions reductions and, 13, 70, 105, 154
equity issues and, 158
fuel costs compared to U.S., 196, 220
natural gas and, 155
population growth and, 106
public participation in policy making and, 143, 149

Federal Energy Management Program, 23
Federal Energy Regulatory Commission, 76
Financial Mechanisms, *see also* Taxes
incentives, 17, 32, 78
investment, 66
pricing policies, 195-201
rebates, 66
Forests
biomass energy and, 17, 94
carbon sinks and, 31, 78
effect of climate change on, 31
tree planting and, 31-32, 77, 97, 105, 144
Fossil Fuels, *see also* Alternative Energy
alternatives to 16-17, 192
demand (projected), 212
former USSR and, 109
oil crisis (1970s) and, 74
Persian Gulf War and, 193
price fluctuations and, 97
price subsidies and, 195-201
reducing consumption of, 189
supply, 161
U.S. dependence on, 211
U.S. sources of, 74
Framework Convention on Climate Change, 208
France, 163, 191-192

GATT, *see* General Agreement on Tariffs and Trade
GDP, *see* Gross National Product
GNP, *see* Gross National Product

General Agreement on Tariffs and Trade (GATT), 24
Geothermal Energy, 192
Germany
balancing economy and environment, 147
natural gas use and, 161
public participation in, 143-144
Global Climate Models, value and limitations, 5
Government (U.S.), *see also* Bush Administration
attitudes of, 170
vs. corporate initiatives, 164
ethics, 169-174, 184
public participation and, 99
Greenhouse Gases, 3-5, 9-10, 13, 61, 70
Green Lights Program, 104
Gross National Product (GNP)
emissions reductions and, 54
energy consumption and, 15, 40, 46, 108, 118, 130
natural resources and, 158
projected growth in, 130

Human Dimensions of Climate Change (U.S. Research Program), 215
Hydroelectric Energy, 17, 192

India, 121, 153
Industrialized Countries
economically developing countries and, 158, 162, 219, 225
behavior changes, 191
Information, *see also* Education
collection of, 10
decision making and, 85
dissemination of, 22, 162
need for, 10, 22-23, 27-28, 89, 115
public education programs and, 85, 117
Inter-Governmental Panel on Climate Change, 153

Japan, 55, 70, 154

Lobbies, and emissions reductions, 43

Manufacturing Sector, 26-28
Market Forces
vs. policies to decrease consumption, 213
vs. regulations, 65-71, 122
vs. technology "forcing," 48
Mass Transit, public attitudes about, 88-89
Media, responsibility for accurate reporting, 115
Methane, 61, 62 95, 108, 155, 159
Montreal Protocol, 158
Motor Vehicle Industry
CAFE standards and, 114
industry perspectives, 111-114
market forces and, 111, 122

Multidisciplinary Approach
in research and policy making, 7
to problem solving, shortcomings of, 183

National Energy Strategy, 67, 73, 74, 79-80, 86-87, 96, 117, 211-213
National Research Council, 193
Native American ethic, 184
Natural Gas
carbon taxes and, 95-96
coal and, 28-29, 155
emissions reductions and, 104
increased consumption of, 107-109, 161, 165
methane emissions and, 95
supply, former USSR and, 161
Natural Resources
in Amazon, 173
international trading of, 158-159
use of, 157-60
world conflicts over, 73, 193, 213
Netherlands, and energy efficiency, 55
New England, emissions reductions in, 165
New England Electric Co., 65-66, 165
Nixon, Richard M. (former U.S. President), energy strategy, 211
No Regrets Policy, 61-63, 121, 161, 164, 190, 199, 224
Nonrenewable Energy, *see* Fossil Fuels
Not-in-My-Back-Yard Syndrome, 92
Nuclear Energy
Chernobyl and, 208
controversial nature of, 189
cost-effectiveness, 208
emissions reductions and, 14, 17, 28, 107, 109, 192-193
European countries and, 192, 218-219
federal subsidies and, 219
improving capacity of n. plants, 121
Pacific Gas & Electric and, 65
reactor safety and, 192
waste disposal, 96, 192, 217, 218
weapons development and, 193

OECD, *see* Organization for Economic Cooperation and Development; *see also* Economically Developing Countries; European Community; Industrialized Countries
Ocean Levels, and global temperatures, 5, 157-158
Office of Technology Assessment, U.S., 14-34
Offsets, *see also* Emissions Reductions; Forests, 78, 92, 116, 154-155
Oil Producing and Exporting Countries (OPEC), 131-140, 155
OPEC, *see* Oil Producing and Exporting Countries

Organization for Economic Cooperation and Development (OECD)
coal consumption and, 109
emissions per capita, 190-191
emissions reductions and, 104-106, 154
Ozone Depletion (Stratospheric), climate change and, 104, 162

Pacific Gas & Electric, efficiency initiatives, 65, 165
Performance Standards, 14, 66
Persian Gulf War
costs of, 213
fossil fuels and, 193
Photovoltaic Energy, as alternative to fossil fuels, 192
Policy Making, *see also* Decision Making, 70, 85-86, 97-98, 107-109, 143-145
Polluter-Pays Principle, vs. emissions trading, 98
Population
climate and, 178-179
growth (projected), 106, 109
emissions increases and, 118
energy consumption and, 127-140
slowed death rates and, 148
policies, 148-49
restraints
economic development and, 147-148, 219, 220-221
education and, 147
need for world-wide, 43
Pricing Reforms, *see* Financial Mechanisms
Public (i.e., the citizenry of the U.S.)
behavior and behavior changes, 15, 42, 81-89, 92-93, 145-146, 169-174, 172-173, 174, 183, 207-208
education
affecting daily choices, 149
behavior changes and, 92-93
for children, 149
effectiveness of, 17, 92-93
need for, 7-8, 209, 221
opinion, on various environmental issues, 81-89, 144-145, 149, 169-174, 212
participation
corporations' need for, 118
in Germany, 143-144, 149
government and, 99
influence on policy making, 143-145
litigation against polluters, 75, 86
Public Utility Holding Company Act, 220
Public Utility Regulatory Policies Act, 212

Quality of Life, and quality of environment, 75

Renewable Energy, *see* Alternative Energy
Reagan, Ronald (former U.S. President), energy strategy, 211
Recycling policies, exemplary public participation and, 144
Regulations, vs. market forces, 65-71
Research and Development
alternative energy and, 25-26, 30, 66-67
collaboration and cooperation needed, 7
current status in U.S., 18
emissions reductions, 14, 17
energy efficiency, 23, 28, 66-67
need for, 18
Responsibility (to protect environment), 7-8, 179-180, 208-209
Rio de Janeiro, *see* UNCED
Risk Assessment, 103, 190

Scandinavia, and energy efficiency, 55
Science, and ethics, 224-228
Scientists' Institute for Public Information, 115
Sinks, *see* Emissions Reductions; Forests; Offsets
Solar Energy, 14, 192, 208, 219
Southern California Edison, 65, 165
Stewardship, and forest use, 32
Strategic Petroleum Reserve (U.S.), 75, 213
Sulfur Dioxide, and cooling effect, 93-94
Sweden
energy costs, 39-40, 196, 198
nuclear energy and, 192, 218-219
Switzerland, 55, 192

Taxes
carbon, 17, 18, 65, 76, 93
analysis of, 50-51, 197-199, 215-216
biomass energy and, 94
coal industry, 200, 216
in EC, 163, 216
economic effects and, 49, 62-63, 92-93, 195, 196, 198, 200
electric utilities and, 29
vs. emissions trading, 89
energy consumption and, 95-96, 213
effectiveness in reducing emissions, 14, 62, 67, 93
fuel switching and, 95-96
GNP and, 63, 198
implementation of, 198-199
manufacturing sector and, 26-27
motor vehicle industry and, 112-113
natural gas costs and, 95
rates (estimated), 66
U.S. resistance to, 143-144
fuel, 24, 97
Gas Guzzler, 24
tax credits, 22, 63, 66
tax incentives, 14, 24, 27
tax rebates, 31, 66, 97, 213
Technology Transfer
for emission reductions, 98, 105, 204
for energy efficiency, 209

Technology Transfer (cont'd)
from industrialized to economically developing countries, 42, 104, 160-162, 219
tax incentives and, 213
Transportation Sector, 23-26, 83-84, 145, 221
Travel Reduction, 16, 23-25, 42, 67, 87, 88, 112-113
Trees, *see* Forests

UNCED, *see* UN Conference on Environment and Development
United Nations (UN), role in monitoring emissions, 218
Conference on Environment and Development, 42, 143, 153, 166, 220, 227
Environment Programme, 153, 220
Fund for Population Activities, 148
United States Government, *see* Government, U.S.
U.S. Census Bureau, 129
U.S. Forest Service, 31

Values, *see* Ethics and Values
Virtual Climate, 203-205, 215-216

Wind Energy, 192
World Bank, 164
World Meteorological Organization, 153
World War II, 224-227

MIX
Papier aus verantwortungsvollen Quellen
Paper from responsible sources
FSC® C105338

If you have any concerns about our products,
you can contact us on
ProductSafety@springernature.com

In case Publisher is established outside the EU,
the EU authorized representative is:
Springer Nature Customer Service Center GmbH
Europaplatz 3, 69115 Heidelberg, Germany

Printed by Libri Plureos GmbH
in Hamburg, Germany